A guide to
choosing a career in
ENGINEERING

(2nd Edition)

Includes a Chapter on

WOMEN IN ENGINEERING

© 2023 **ChuDace Publishing**

ISBN: 978-0-9985843-93 [Second Edition]

Published and printed in the United States of America

Preface

The *Engineer* is the chair of a technology trio who create innovations that complement or replace human effort, and enhance human development. The *Technician* is the artisan who transforms the Engineer's design sketches and calculations into working drawings and, ultimately into products that meet human needs, under the management and supervision of the *Technologist*. The first edition of this book (2019) discussed extensively the distinctive skills that characterize each technology specialty, the interdependence and complementarities of the three different skill sets, the prime role of the engineer as team leader, and the type of training required to produce a professional engineer in the main areas of specialization. This edition extends the scope of specializations to include several new and emerging areas. Engineering is gender-neutral but various studies have shown that there is a glaring paucity of women in STEM and engineering professions all over the world. The first edition discussed this issue briefly, but this edition includes a chapter that presents an in-depth discussion in an attempt to establish the extent of the problem, identify the issues, and propose possible solutions.

Engineers are creators and motivators of major technology innovations: they transform the natural resources and materials in the world around them to create new technologies that complement human effort; they design products and production processes; they design and optimize machinery and structures; they manage production; they drive innovation. It is difficult to imagine an area of human development that is not positively impacted by the products of engineering: building, manufacturing, machinery, energy, agriculture, water supply, appliances, consumer products, transportation, communication, health services, education, environmental health, entertainment, and many more. Throughout history, technology and society have co-evolved and, today, the symbiosis is such that societies shape technologies, and technologies also shape societies. Technology innovations are driven by societal needs and wants, while many new and emerging technologies create new needs and wants.

Human development has benefited tremendously from technology but the negative impacts and unanticipated consequences on society and the environment are becoming increasingly apparent as technology becomes more pervasive or powerful. All technology innovations come with consequences and trade-offs many of which are having increasingly negative impact on society and raising significant concerns, notably security/safety, environmental and ethical issues. It is vital therefore that engineers who are the main creators and prime movers of technology innovations should also be aware of the potentially negative impacts on society, especially because many will rise to management positions and have to take vital technological and ethical decisions which require making informed choices between competing priorities. This edition includes a new chapter on The Engineer and Society in which some of the negative consequences of technology innovations on the environment and society are discussed in some depth. Urgent issues discussed include the increasingly severe environmental damage and associated consequences resulting from uncontrolled energy production and use, and socio-ethical issues arising from many information and other technology innovations, notably: uncontrolled invasion of privacy, digital piracy, increasing deployment of automation, and proliferation of Artificial Intelligence in many areas of societal life. Hopefully the discussions initiated in the two new chapters will help to energize further discussions and actions that promote awareness and eventually lead to solutions that enhance human development.

Adeniyi A. Afonja
Professor Emeritus
Materials Science & Engineering
& Materials-Energy-Environment consultant

The Engineering Profession

Engineering is one of the oldest professions, taking its roots from the documented common practices of the Early people dating back some twelve thousand years. Science evolved from the Renaissance era and seeks to understand the natural environment, engineering seeks to shape it through the design, manufacture and maintenance of technological systems, creating a very wide range of products that propel human development: from non-stick cooking pans, through digital smart cars and homes, to space rockets. Engineering draws heavily from a very wide range of complex bodies of knowledge - science, mathematics, social sciences, humanities, empirical evidence - to create the complex products and infrastructures that currently dominate the human environment. Engineering shapes our lifestyle and culture: it is difficult to imagine a minute in our daily lives when we are not in touch with a product of engineering. Starting from the humble beginnings of creating objects and procedures that make life easier on earth, the scope of engineering has expanded to include softer issues: aesthetics, safety, and ethical issues on how its products impact on, influence and serve social values. The negative impacts of engineering on human life and the environment are becoming increasingly topical and the focus of engineering has expanded beyond the immediate use of a product to the total *lifeycle* of the product: where the raw materials will come from; how they will be produced; the conditions under which they will be used; and how they will be disposed of at the end of life, all of which could have very significant negative impacts on both life and the environment. The profession has also moved from core hardware to soft areas including software, bio-, genetic and artificial intelligence engineering.

Engineering is a prime profession of the past, present and future and, in any country, engineering professions typically have job prospects that pay a wage higher than the national median. University graduates in engineering can expect to earn a starting salary that is at least 20% more than the average graduate salary, and a lot more during their career when they are expected to have risen to management positions in major engineering enterprises. Various recent studies project that the oldest and most versatile specializations in civil and mechanical engineering will continue to dominate employment opportunities in the profession, accounting for around 60% of new jobs for engineers worldwide. Around 30% will be shared by the newer sub-disciplines, notably electrical, materials, chemical and computer engineering, while the highly specialized ones such as computer, telecommunications, bio-, cyber- engineering will share the balance.

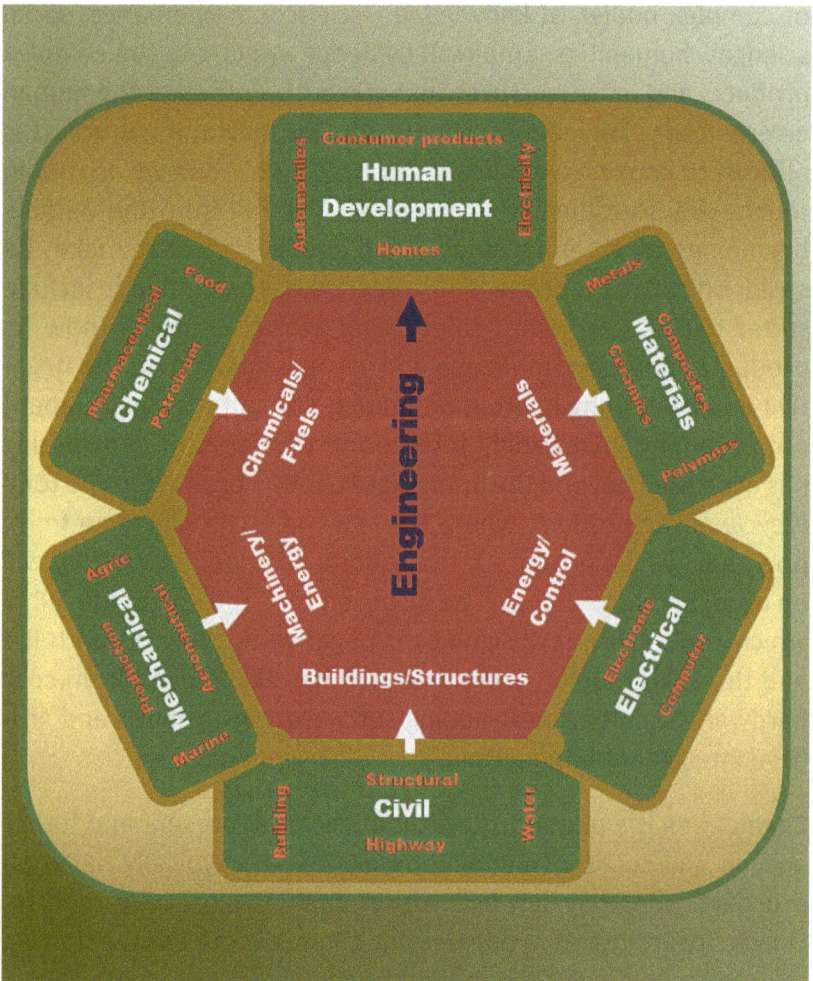

About the author

Adeniyi Afonja graduated in mechanical engineering from King's College, University of Durham, United Kingdom in 1965. He obtained a masters degree in metallurgical engineering and management in 1966 and doctorate in mechanical engineering in 1969, both from University of Aston, Birmingham, also in the United Kingdom. He has over fifty years of cumulative experience in engineering education and senior administrative positions at various universities, and also in engineering consulting. He has held many visiting engineering professorship and fellowship positions in international institutions and research establishments, including Massachusetts Institute of Technology, University of Wisconsin-Whitewater, both in the United States; Energy Research laboratories, National Research Centre (CANMET), Ottawa, Canada; and University of Newcastle Upon Tyne, Energy Research Centre, United Kingdom. Professor Afonja has won many international fellowships including the Commonwealth Fellowship, and Senior Commonwealth Fellowship (United Kingdom); Fulbright Fellowship, and Senior Fulbright Fellowship (United States); Research Fellowship (Canada), United Nations Industrial Development Fellowship, Ukraine. He retired in 2005 and was appointed Professor Emeritus, Department of Materials Science and Engineering, OAU, Ile-Ife, Nigeria in 2006. He is still active in materials-energy-environment research and consulting, and has published many books in all the areas in recent years. He has also published a book on Historical Milestones in Engineering and Technology, and co-authored another on Cyber-security.

Professor Afonja is a Chartered Engineer (United Kingdom), Registered Engineer (Nigeria), Fellow of the Institute of Materials (United Kingdom and Nigeria). He is also a Fellow of many national and international academic and engineering societies, and has served as consultant on numerous industrial engineering and energy projects. He is married and blessed with children and grandchildren.

Engineering at work: Burj Khalifa, Dubai, United Arab Emirates, one of the world's tallest buildings. *(popularmechanics.com)*

Table of contents

Complex bridge network
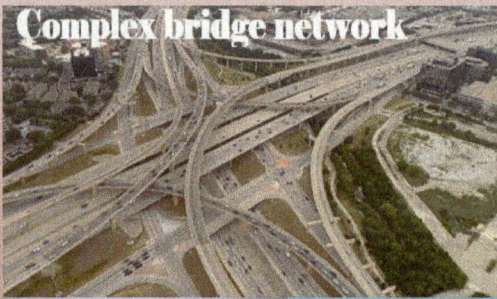

Automobiles

Sky scrapers

Ships

Trains

Airplanes
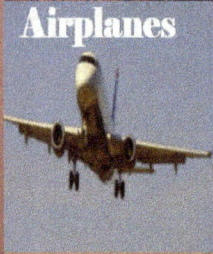

Space rockets
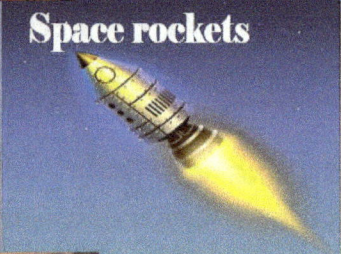

Military weapons

Agric machinery
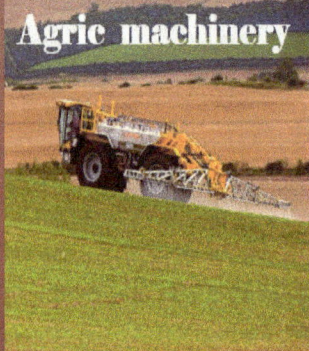

Construction machinery
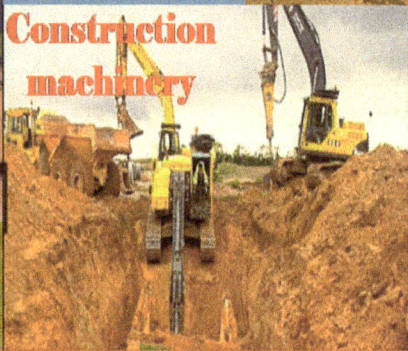

Steel products
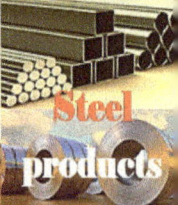

Consumer products

PRODUCTS OF ENGINEERING

(Google images)

Inspirational Quotes about Engineering I

"The story of civilization is, in a sense, the story of engineering - that long and arduous struggle to make the forces of nature work for man's good." – Lyon Sprague DeCamp

"The well being of the world largely depends upon the work of the engineer. There is a great future and unlimited scope for the profession; new works of all kinds are and will be required in every country, and for a young man of imagination and keen-ness I cannot conceive a more attractive profession. Imagination is necessary as well as scientific knowledge."
–Sir William Halcrow

"It is a great profession. There is the satisfaction of watching a figment of the imagination emerge through the aid of science to a plan on paper. Then it moves to realization in stone or metal or energy. Then it brings jobs and homes to men. Then it elevates the standards of living and adds to the comforts of life. That is the engineer's high privilege." – Herbert Hoover

"The ideal engineer is a composite... He is not a scientist, he is not a mathematician, he is not a sociologist or a writer, but he may use the knowledge and techniques of any or all of these disciplines in solving engineering problems." –N.W Dougherty

"As engineers, we were going to be in a position to change the world – not just study it." –Henry Petroski

"The civil engineer builds the world; the computer/electronic/telecom engineer connects it; the electrical engineer powers it; the mechanical engineer moves and feeds it; the chemical engineer keeps it alive and healthy; the materials engineer empowers all engineers". – Adeniyi Afonja

Inspirational Quotes about Engineering II

"All we know about the new economic world tells us that nations which train engineers will prevail over those which train lawyers. No nation has ever sued its way to greatness." – Richard Lamm

"A good scientist is a person with original ideas. A good engineer is a person who makes a design that works with as few ideas as possible. There are no prima donnas in engineering."
— Freeman Dyson

"I am, and ever will be, a white-socks, pocket-protector, nerdy engineer, born under the second law of thermodynamics, steeped in steam tables, in love with free-body diagrams, transformed by Laplace and propelled by compressible flow."
– Astronaut Neil Armstrong

"Design is not how it looks like and feels like. Design is how it works" – Steve Jobs

"At its heart engineering is about using science to find creative practical solutions. It's a noble profession."
– Queen Elizabeth II

"Science is about knowing, engineering is about doing."
– Henry Petroski

"Engineering is the art of modeling materials we do not wholly understand, into shapes we cannot precisely analyse so as to withstand forces we cannot properly assess, in such a way that the public has no reason to suspect the extent of our ignorance."
— AR Dykes

"You never change things by fighting the existing reality. To change something, build a new model that makes the existing model obsolete" – R. Buckminster Fuller

Inspirational Quotes about Engineering III

"Strive for perfection in everything you do. Take the best that exists and make it better. When it does not exist, design it".
– Sir Henry Royce

"Science can amuse and fascinate us all, but it is engineering that changes the world." – Isaac Asimov

"The way to succeed is to double your failure rate."
– Thomas J. Watson

"The scientist discovers a new type of material or energy and the engineer discovers a new use for it."
– Gordon Lindsay Glegg

"We are continually faced by great opportunities brilliantly disguised as insoluble problems."– Lee Iacocca

"Scientists investigate that which already is; Engineers create that which has never been."— Albert Einstein

"Engineering is the art of directing the great sources of power in nature for the use and convenience of man."
–Thomas Tredgold

"What we usually consider as impossible are simply engineering problems... there's no law of physics preventing them."
– Michio Kaku

"Engineering is the professional and systematic application of science to the efficient utilisation of natural resources to produce wealth." — Theodore Jesse Hoover

"Left to the engineer, the product needs to be perfect, but it never will be." – Adeniyi Afonja

Inspirational Quotes about Engineering IV

"Failure is the opportunity to begin again more intelligently."
– Henry Ford

"Engineering is not merely knowing and being knowledgeable, like a walking encyclopedia; engineering is not merely analysis; engineering is not merely the possession of the capacity to get elegant solutions to non-existent engineering problems; engineering is practicing the art of the organising forces of technological change... Engineers operate at the interface between science and society." — *Gordon Stanley Brown*

"The life work of the engineer consists in the systematic application of natural forces and the systematic development of natural resources in the service of man."
— *Harry Walter Tyler*

"The engineer is a mediator between the philosopher and the working mechanic and, like an interpreter between two foreigners must understand the language of both, hence the absolute necessity of possessing both practical and theoretical knowledge." – *Henry Palmer*

"Engineering problems are under-defined, there are many solutions, good, bad and indifferent. The art is to arrive at a good solution. This is a creative activity, involving imagination, intuition and deliberate choice." – *Ove Arup*

"Scientists study the world as it is; engineers create the world that has never been." –*Theodore von Karman*

"I have not failed. I've just found 10,000 ways that won't work....Genius is one percent inspiration and ninety percent perspiration". – *Thomas Edison*

1   What is Technology?

1.1   INTRODUCTION

Technology is as old as mankind: the Early man had to develop different tools and devices for survival, comfort and as a complement to inadequate human effort. Shelter was perhaps one of the most urgent needs which was solved initially by digging caves on mountain sides (excavation/tunneling in modern civil engineering terminology). It did not take long to develop technologies for construction of huts with molded clay walls and thatched roofs, the humble beginnings of civil engineering and construction. It is interesting to note that the Early human was well aware that mixing the clay with straw improved strength significantly, the humble beginning of modern composite materials engineering. Hunting was a major problem since most animals can run faster than humans. This led to the invention of traps, catapult, slings, bows and arrows (energy conversion in modern mechanical engineering terminology), which was made possible by the inventions of hand tools (carving tools from stone and forging knives, arrow tips and other components from meteoritic iron): the humble beginning of mechanical engineering. The early human soon learnt the art of casting tools and ornaments from ferrous and non-ferrous metals and alloys produced by smelting ores with charcoal and coal outcrops (the humble beginnings of metallurgical engineering). Some of the processes they used (such as chipping, cold and hot forging, drawing, lost-wax casting, alloying) still feature prominently in modern mechanical and metallurgical engineering production processes. Over the centuries, technology has evolved through many development   phases, but the focus has remained unchanged: development of processes and products that meet human needs, provide comfort, and complement human effort.

One of the most dramatic milestones in technology was the beginning of industrial mass production towards the end of the seventeenth century which was made possible by availability of coal in commercial quantities, which in turn was a result of the development of technologies for recovering underground mineral deposits (early mining and minerals engineering). The impact of this development on virtually all aspects of human existence was dramatic: it nucleated the Industrial Revolution; it made possible mass production of metals, chemicals, pharmaceuticals; it led to the invention of the steam engine, railway, mechanized agriculture and industrial production; it facilitated the development of the automobile, internal combustion engines, the energy industry, and many more.

1.2   ENGINEERING AND TECHNOLOGY

Technology may be defined in many ways but, basically, it is the art of manipulating the natural environment to create complements to human effort and competence, thereby promoting and enhancing human development: shelter, tools, machines, energy, drugs, consumer goods, transportation and communication technologies, artificial intelligence, etc. Most of the early technological innovations were driven by intuition, inquisitiveness, trial-and-error techniques, and perseverance. However, developments in science that have defined and characterized the natural environment in the last five centuries or so have transformed technology into a profession based on a logical scientific thought process. Modern technology includes all types of human-created systems and processes that result from practicalizing scientific theories on the natural environment. The products of technology may be physical, virtual, or intellectual; and creation, operation and maintenance require a diversity of skills. Engineers have the skill to design products from first scientific principles: materials, machinery, structures, processes, products, engineering systems, etc. They also form the core of engineering education and research. Engineering technologists have similar status and prospects as engineers in that both require college degrees but their training focuses on operation of industrial/production engineering processes that transform the engineer's design into products. These include manufacturing process planning and operation, quality assurance, machinery installation and maintenance.

Engineers require strong grounding in physical, chemical, and biochemical sciences, the relative emphasis depending on the specialization. Civil, mechanical, electrical, computer engineering and sub-specialties require strong skills in the physical sciences; bioengineering students need strong skills in biochemical sciences; while those in chemical, and materials engineering require strong skills in both areas. The training of technologists is similar to that of engineers in corresponding specializations but the emphasis is on applications and operations rather than on design and mathematical analysis. Engineering technicians are the artisans who install, produce, operate, and maintain all creations of technology. They do not require college degrees and acquire specific skills such as machining, welding, fitting, air conditioning, from polytechnics, trade schools, on-the-job training, or junior colleges. In effect, engineers are a prominent part of the family of technologists, and the different categories of skills are both complimentary and indispensable in actualizing technology innovations. Engineers are the major creators of technologies and innovations, engineering technologists (better known as technologists) are the production planners and managers, and engineering technicians are the hands-on doers.

1.3   ROLE OF SCIENCE IN ENGINEERING AND TECHNOLOGY

Technology has evolved from Early times when it was a practical think-it-do-it process, stimulated largely by the need to solve pressing problems, but also in response to intuition. For example many of the early technologies targeted provision of adequate and effective shelter and development of tools for farming, hunting and warfare. However, the development of the steam engine, considered the most important invention and prime mover of the Industrial Revolution era from the late 1600 AD was a result of intuition believed to have been inspired by an observation of the way in which boiling water moved the lid of a kettle: it should be possible to harness steam energy to move other things, like grain mills, pumps, trains, or even the first steam-powered vehicles. The origin of this intuition is unclear but many inventors had designed primitive steam engines from around the first century AD. However, the real breakthrough came in the seventeenth century when the first practical steam engines emerged, used for pumping water out of flooded coal mines in England. This made possible the mass production of coal which powered the Industrial Revolution.

Virtually every early technological invention emerged through a long series of trials, errors, failures, even fatal accidents until useful products and processes were achieved. The plight of early inventors is aptly summarized in the words of Thomas Edison, one of the world's most prolific early inventors, credited with the invention and commercialization of many devices in fields such as electric power generation and distribution, and invention of many technologies including the electric bulb, mass communication, sound recording, devices, and motion pictures:

> *" I have not failed. I've just found 10,000 ways that won't work.........Genius is one percent inspiration and ninety-nine percent perspiration".*

Edison had very little education and only attended school for a few months. He was largely self-educated by reading scientific textbooks and taking advantage of The Cooper Union which offered tuition-free courses to talents that might otherwise have gone undiscovered.

Science is a logical thought process that studies the natural environment and seeks to explain well-established technology observations, inventions and processes. Archimedes who lived in the sixth century BC was probably one of the first to appreciate the powerful relationship between science and technology and is widely regarded as one of the pioneers of the evolution of engineering as a profession. He used the principles of mathematics and physics to explain observed technology phenomena and devised scientific experiments which led to the logical and theory-based design of a wide range of inventions including mirrors capable of concentrating the Sun's energy (a fundamental basis for modern solar energy). He was a prolific inventor and many of his inventions still feature prominently in modern engineering, notably the cantilever system and the screw pump/propeller (Figure 1.1). Another major and enduring contribution of Archimedes was in the field of fluid mechanics, a core subject in modern basic engineering studies.

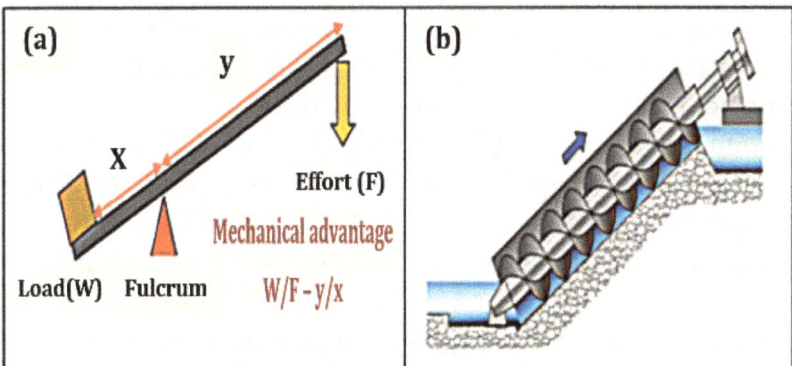

Figure 1.1. (a) Archimedes cantilever law; (b) Archimedes screw pump [*(b) from ksb.com*]

For example, the early humans used boats and canoes developed by trial and error but Archimedes sought to explain through an intensive logical process, why a boat floats and under what conditions it can remain afloat. This led to the development of a series of equations known as *Archimedes Principle (the buoyant force on a submerged object is equal to the weight of the fluid that is displaced by the object)* which has remained a major fundamental concept in fluid mechanics, marine engineering and modern ship design until today.

Since Archimedes, many prominent scientists have developed scientific theories to back up practical observations, notably English Sir Isaac Newton's three laws of motion and the three laws of thermodynamics which evolved from the work of different scientists over most of the nineteenth century. Over time, science has evolved as a major trigger for technology development, virtually eliminating the traditional strategy of trial and error. For example, generation of electricity is a natural phenomenon and occurs frequently in thunderstorms. Although the early people were apparently aware of the potential usefulness of harnessing this energy source (and there is ample archaeological evidence that copper-iron-based batteries were developed in the ancient Middle East several hundred years BC), it took the work of many prominent scientists culminating in the monumental work of English Michael Faraday who studied the basic science behind electricity and developed the first electric dynamo in 1831. This fundamental work made possible the continuous generation of electricity and provided the basis for subsequent work by many engineers including Americans Edison and Nikola Tesla, to move electricity through different iterative stages to its present indispensable human development utility.

Most of the major developments in electrical, electronic and nuclear, materials science and engineering in the last two hundred years or so have been inspired by well-developed scientific theories and principles. German-American Albert

Einstein, the theoretical physicist is probably the most famous scientist of the 20th century. His works on the theory of relativity, quantum mechanics, and his famous mass-energy equivalence formula ($E = mc^2$) for which he won the Nobel Prize for Physics in 1921 have led to major technological developments including nuclear energy, theory of gravity, the photon theory of light, and advances in astrophysics. The invention of the transistor in 1948 resulted from the work of three American physicists, William Shockley, John Bardeen and Walter Brattain for which they shared the 1956 Nobel Prize for physics. The transistor is considered one of the most important inventions in history. It led to the development of light-current electronics and production of integrated circuits (chips) which miniaturized everything from computers to telecommunication devices and equipment, control systems, appliances, medical implants, etc.

Engineering is the branch of technology that involves the application of fundamental laws of science and logical thinking to the design of products and processes, thereby largely eliminating the time-consuming, wasteful and dangerous trial-and-error approach. For example, the history of aviation dates back more than two thousand years, when various inventors attempted to emulate the flying bird by producing kites, gliders, balloons, etcetera, and there were many fatalities. However, the monumental work of the Wright brothers in the early part of the twentieth century was the first major attempt to combine the principles of science and technology in producing the first prototype that eventually evolved into the modern aircraft. Although the first flight lasted only 12 seconds, rose to a height of 6 meters and covered only 40 meters, it was the first successful powered flight whose design was based on scientific principles and the success quickly led to the design of the Wright flyer (Figure 1.2). This life-changing development would not have been possible without the pioneering work of Sir Isaac Newton (1643-1727) on motion, the work of Sir George Cayley (1773-1857) on the physics of flight; the work of Samuel Langley (1834-1906) 'Experiments in Aerodynamics';

the Laws of Thermodynamics (which emerged from the work of many scientists); and the contributions of many other scientists. It is interesting to note that, like many other early inventors, the Wright brothers had no formal science education: they were bicycle builders and repairers. However, they self-educated by reading numerous scientific publications, they complemented with their experience in mechanical workshop practice, and succeeded in resolving aviation problems which had confounded solution for centuries.

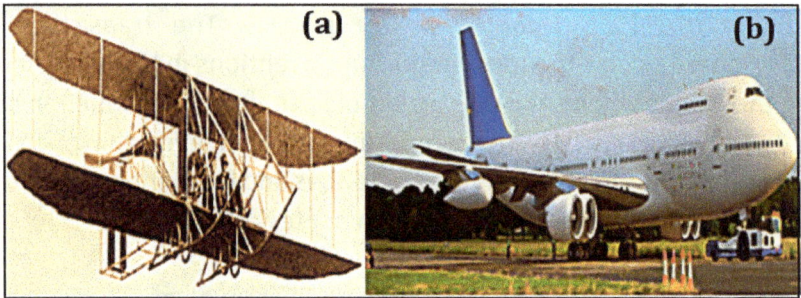

Figure 1.2. (a) The Wright flyer, the first successful manned and engine-powered flight in 1903 (b) A Boeing 747, 1974. *(Southernboating.com; aviationfilming.com).*

Since the first successful manned flight, the systematic approach to design has evolved to the point that the biggest airplanes are now designed from first principles with such precision that all the thousands of components can be replicated in mass production, and the first test flight is largely successful. The modern internal combustion engine which powers everything from automobiles, space rockets, to gardening equipment is another revolution made possible by technologists seeking to practicalize scientific principles of fuel combustion. Electricity was the product of atomic physics and materials science, and many other inventions that currently enhance human development were inspired by scientific principles: machinery, refrigeration and air conditioning, telecommunication gadgets, computers, cell phones, remote controls, new materials, biomedical implants, pharmaceutical products, and consumer products.

1.4 THE TECHNOLOGY FAMILY

Technology derives from the word 'technique' (which means a way of carrying out a particular task or process), and is applicable to most areas of human endeavor. However, in the current context, it may be defined as a successful development and use of techniques for augmenting human effort and producing goods and services. In effect technology is a synthesis of processes which include conception, development, production, operation, maintenance, and end-of-life disposal. Although most of the stages were usually carried out by the inventor in earlier times, modern technology involves different levels of expertise provided by the engineer, the technologist and the technician (artisan/operative), all working together to achieve a desired goal (Figure 1.3). There are three major specialists in the technology team: the engineer, the engineering technologist (better known as technologist) and the artisan/operative, known collectively as technician. However, involvement in a particular project depends on the nature and complexity.

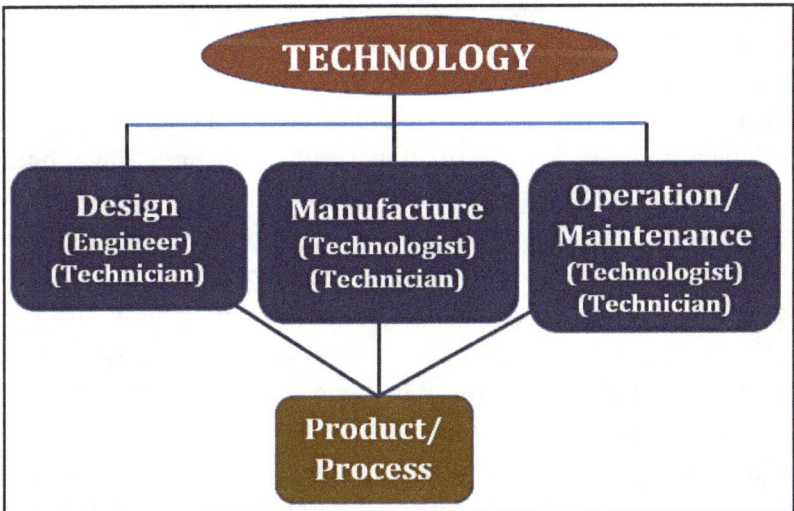

Figure 1.3. Stages of product/process development and the required expertise.

Engineers feature primarily in product and process design and development, technologists specialize in equipment installation, production and maintenance management, while technicians are usually involved in both. All three skill levels feature prominently in all major areas of technology (Figure 1.4).

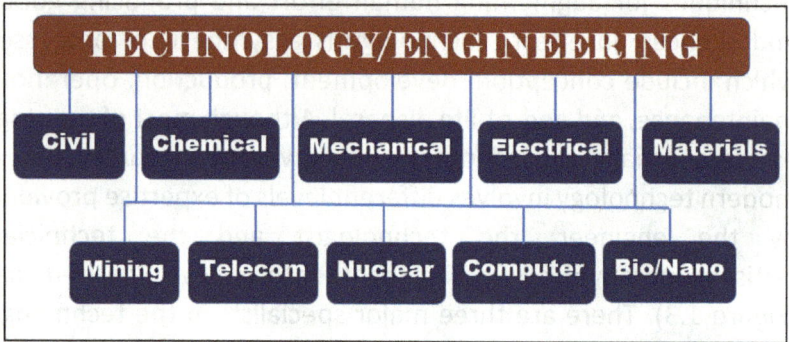

Figure 1.4.   Main types of technology (every specialty features engineers, technologists and technicians).

Every specialty in technology features the three levels of competence but the complexity of the production/process technology determines the skills and competence required in specific aspects. For example, established production operations in which the products, processes and machinery have been externally designed may be coordinated and supervised by technologists but those that involve new products, processes or significant alterations may require the expertise of engineers. Service and repair shops and installations may be supervised by technologists but most of the actual work is carried out by technicians. It should be noted however that although each level of skills is critical in engineering and technology, the public is largely unaware of the different levels of technology expertise, and, as far as they are concerned, technologists and technicians are also engineers. In fact, they are more likely to interact with them than engineers in their daily lives.

1.4.1   The Engineer

The engineer plays a leading role in the design and development of new products and production processes, and needs strong competence in scientific and basic engineering principles to perform. Engineering training assumes a strong grounding in the physics and chemistry of atoms, the basic building blocks of matter (solid, liquid, gaseous, plasma); the fundamental laws of physics and chemistry; and mathematics which is the indispensable analytical tool. This fundamental knowledge facilitates an understanding of the structure, properties and behavior of masses (solid materials, chemicals, gases, liquids) that most engineers have to work with irrespective of specialization. The engineer is primarily responsible for the design stage of a product or process, but also often manages operation and maintenance depending on the size and complexity of the setup.

The engineer is trained to develop and perfect a design, largely by asking questions and seeking answers, mostly based on scientific principles. Take for example the design of a simple bolt and nut fastener, one of the most versatile engineering components that is found on almost any mechanical or household product. The engineer first gathers information on what the product will hold together; the types and intensities of stresses to which it will be subjected (tensile, compressive, bending, static or dynamic load, or a combination of several stress types); the nature of the environment in which the component will operate (hot, cold, freezing, corrosive or a complex combination); aesthetic requirements (does it need to be visually attractive?); and acceptable cost of the product. On the basis of the principles of stress analysis (civil/mechanical engineering), the dimensions that can bear the load are determined; the engineer then chooses the appropriate metal or alloy that can withstand the types of stresses in the specific environment, guided by principles of materials science and engineering. The appropriate production method is chosen depending on the level of service integrity required. While machining may be adequate for many applications, forging may be required where fatigue failure resistance is important. Furthermore the product may need to be plated to elevate oxidation/corrosion resistance, and/or polished if it needs to be visually attractive.

The engineer also provides an estimate of the likely cost of production and, finally, the design is presented in form of sketches, flowcharts and calculations for the technician (draftsman/draughtsman) who develops working drawings that will guide the technician (artisan/operative) who will make the product, supervised by the technologist. The same systematic procedure applies to very complex projects like industrial machinery, jet engines, aerospace rockets and associated equipment, bridges, complex road networks or skyscraper buildings, all of which are broken down to simple components for eventual assembly. For example, the average automobile comprises around two thousand components, each designed in accordance with the logical thought process outlined above. The engineer or technologist or both may be involved in supervision at the production and maintenance levels again depending on the complexity of operations.

1.4.2   The Technologist

Technologists have virtually the same scope for specialization as engineers but their main competence is in managing production and maintenance. They are not trained to be designers but, with experience, they can become competent in this area as well. It should be noted also that most of the early technology inventors had no formal training in engineering: they relied almost entirely on intuition, trial and error, and self-education. Technologists can manage production systems that operate on established designs and procedures, mechanical workshops, building projects, telecommunication stations, etc. They also feature prominently in technical sales and marketing, equipment installation and commissioning. Many technologists acquire further training and end up as engineers, or, through experience, can rise to the top in industry.

1.4.3   The Technician

Artisans (skilled in a specific trade, craft or operation) are collectively known as technicians. They are in the lowest rank of the technology profession, yet they are the real doers and are indispensable in virtually all aspects of technology. They are the draftsmen who translate the design of the engineer into working drawings for production; they are the equipment installers; they are the industrial producers (machinists, fitters, welders, plumbers, air conditioning technicians); they maintain industrial and domestic equipment; they operate and maintain the numerous instruments and equipment that facilitate the work of industry, health, education, research and development. They also feature prominently in technical sales and services. In fact, they are the ones known by the general public as engineers.

1.5   TECHNOLOGY TRAINING

The typical engineering degree program provides a strong foundation in science, mathematics, basic engineering science (engineering mechanics, thermo-fluids, materials science, basic electrical engineering, engineering drawing). Students also learn the importance of ethics, safety, professionalism, and socio-economic concerns in resolving technical problems through synthesis, planning, and design. The engineer would have successfully completed a 3-5 year college degree program in one of the many specialties, with emphasis on engineering analysis and design. The first year of a typical engineering program (two in some cases) deals with the application of basic-physico-chemical principles to practical situations: mechanics of machines, materials and fluids; thermodynamics of solid materials and fluids; engineering materials; electronic and electrical properties of atoms; and engineering drawing. The last two years (one in some programs) focus on   topics and applications that are specific to each specialization. For example, mechanical engineering training will now involve in-depth analysis and design of machines, heat engines, automotive, aero, agricultural and marine mechanical structures, power generation (production of heat, steam, compressed gases), boilers, furnaces, heat exchangers, turbines, etc., while chemical engineering focuses on transport processes and unit operations (fluid, heat and mass transfer, filtration, mixing, chemical reaction engineering, catalysis, etc.).

Civil engineering programs will focus on structural design, building design, building materials, construction. Throughout the training, the student is exposed to a lot of complementary laboratory and design studio work, and industrial attachment. Also, most students have final year projects designed to test their overall knowledge of the specific specialization in engineering. Most engineering degree programs also include complementary subjects such as economics, management, sociology, engineer and society. By the end of training, and with appropriate pupilage (a 2-year post-graduation industrial experience), the engineer is well equipped for a career in product/process design, and management of production or maintenance.

Admission requirements for an engineering degree program vary very widely: many institutions admit high school diploma, school certificate, general certificate (ordinary level) graduates who spend the first year on basic sciences before proceeding to the 3-year or 4-year degree program while others only admit students with advanced level grades in the basic science subjects, or college diploma in relevant subjects. Most institutions arrange for their engineering students to spend at least two end-of-session vacations or one full year in industrial attachment before graduation. In some cases, candidates spend two years each before and after a one-year industrial attachment and are awarded a master's degree. All fresh graduates undergo a period of apprenticeship usually for two years before they qualify to practice as registered/chartered engineers. Some registration systems are more specific, such as certified structural, building, highway, water engineer (sub-specialties of civil engineering), or mining engineer, industrial engineer, production engineer, automobile engineer, marine engineer, aeronautical engineer, aerospace engineer, robotics engineer, mechatronics engineer, etcetera (all specializations in mechanical engineering).

The technologist has a similar college training as the engineer and, although course structure is usually similar, the emphasis is less on theory/design and more on competence in operations and maintenance. Technicians are normally trained on a diploma program in a polytechnic, college of technology, technical college or trade school, but many are also trained on the job, complemented with evening or on-line classes. While they will learn some fundamental principles of engineering specific to their areas of specialization, the emphasis is on acquiring practical skill in a chosen specialization: machining, welding, fitting, electronic and electrical work, carpentry, telecommunications, instruments technology, draftsmanship, refrigeration and air conditioning, bricklaying, plumbing, auto repair, etc. Many institutions now offer continuing education pathways for artisans to upgrade and eventually become technologists or engineers. Like engineers, technologists and technicians also have registration bodies.

1.6 ENGINEERING DESIGN

As mentioned earlier, design competence is the primary qualification that sets the engineer apart from other technologists and the training program provides all the analytical tools required for the various stages of the design process. The design of an engineering product evolves from a systematic, logical thought process, from conception to finished product; it may involve hardware, software, or a process which may be a combination of hardware and software. Most design processes require the application of fundamental principles of science and mathematics, a good working knowledge of engineering materials, manufacturing processes and engineering economics. Some design projects may also require the input of several different engineering specialties. For example, it is not unusual to complete a design only to find that appropriate materials are not available, or the design is too complex to be produced economically. In effect, most designs require the input of different engineering skill sets. It is vital therefore that the engineer has a holistic awareness of the basic steps involved in design, the skill inputs required, the potential pitfalls, and the techno-economic viability of the product.

1.6.1   The Engineering Design Process.

The Engineering Design process involves six clear steps as shown in Figure 1.5. The product no matter how small requires all steps in order to meet performance requirements. The average automobile has around two thousand components while the aircraft has more than twelve thousand, and each component has to be properly designed, manufactured, and interfaced with other components. Some designs may be completed in a couple of days while others like airplanes may take years before they are perfected. It should be noted however than once a design is finalized, it forms the basis for replication by production until a modification is desired. It should be noted also that the design of processes and software follows similar procedures.

1.6.1.1   *Problem/Need Definition*

The need for the design of a product may emanate from intuition, the desire to solve a problem, or the need to improve an existing product or process. For example, prior to the development of the first smart phone, the average executive on a trip carried a clumsy box load of expensive gadgets including a cell phone, a CD player, a CD caddy, a camera, a tape recorder, a portable scanner, a laptop. Clearly there was a problem here which remained latent until Steve Jobs-led Apple Corporation found a way of combining most of the functions into a small, portable, affordable device, through a long series of prototype design, testing and improvements to arrive at the first *i*Phone which has evolved into today's smart phone. The need had been evident for years but it needed intuition to trigger action, and persistence to succeed, two hallmarks of the entire life of Steve Jobs. Many other inventions are not necessarily addressing a problem but making life much easier or comfortable, such as the ubiquitous appliance remote control or the numerous appliances and accessories that now feature in the modern home or automobile; smart digital health monitors; self-driving automobiles; voice control of devices, equipment, and other everyday-use technologies; or the thousands of new software applications that appear for every conceivable use by the day.

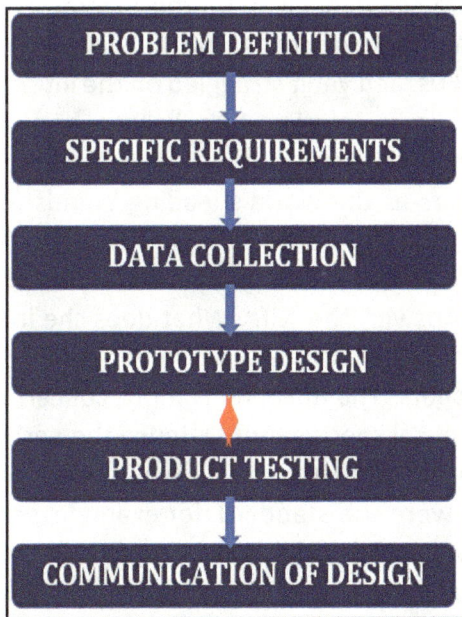

Figure 1.5. Main stages of the Engineering Design Process.

Virtually every product design process starts with the recognition of a problem, a need, the desire to create a new product or make an existing one better. For example, an existing product may be considered deficient but quite often the users are unaware that there could be a better way of doing things or that an existing product could be improved in terms of function and aesthetics. Also, the unexpected appearance of a new invention or innovation can trigger 'want'. The smart watch, *i*Pod, *i*Pad are typical examples. This is where the inventor, innovator, or entrepreneur thrives: producing the unexpected and promoting en-mass acceptance.

1.6.1.2 *Specific Requirements*

Market research is very useful in identifying the potential needs and wants of consumers, to test the potential reaction of consumers to the planned product, and to identify potential competition. It is not unusual to produce a technologically sound product only to see it struggle on the market. For example, the British Chieftain tank was recognized as the world's leading battle tank in the 1960s, and yet it struggled on the international market against technically inferior but lighter and cheaper tanks produced in other countries. Another example is the emergence of Japanese cars as the world's leading brands from the 1960s, thanks to market research which provided answers to two key questions: whose input and opinion matter more in the choice of a family car? Answer: the wife. What does she look for in a car? Answer: aesthetics, upholstery, air conditioning, complementary electronic gadgets. The husband is more concerned with under-the-hood technical specifications. Hence the early Japanese cars were designed to attract the wife while many of the technical specifications were sub-standard, for example, using drum brake systems when the world had already moved on to much more efficient, safer but more expensive disc brakes. Also, the average consumer is now more prosperous and tends to change consumer products frequently, hence manufacturers have largely abandoned durability in favor of aesthetic appeal.

Once a problem, need or potential want is identified, the key questions are, what is the product required to do? Who will be the primary user? What other products will it work with? In what kind of environment will it operate (temperature, stresses, friction, corrosion, erosion, etc.)? What are the most important features of the proposed product? What types of materials can withstand the specific environment? What existing products will it replace or compete with? Which manufacturing processes will be required? How critical is safety? How important is aesthetics or cost?

1.6.1.3   *Data Collection*

Once the requirements are established, the designer conducts research and data collection on existing similar products and designs, potentially suitable materials, information on other components that the product must integrate with, possible production processes, etc. Some laboratory investigation may be necessary when appropriate data is not available. If for example there is inadequate performance data on the potential materials or the product will come in contact with a chemical for which corrosion data is not available, laboratory testing may be necessary.

1.6.1.4   *Prototype Solution*

On the basis of the data collected, potentially suitable materials are selected and necessary calculations are carried out to determine appropriate dimensions and develop a prototype design. Necessary working drawings are produced, the manufacturing and assembly flowcharts are designed and a prototype is produced. The invention of the computer has greatly simplified the engineering design process: the complex calculations which could have taken months to complete by traditional methods can now be carried out in minutes and there are now many design software packages for all sorts of applications. Furthermore, there is extensive technical information on a wide variety of materials, material properties in different environments, and selection criteria.

1.6.1.5   *Product Evaluation*

The prototype is tested to determine the extent to which it meets the desired specifications and performance requirements. The potential cost of the product is also determined. Hardly would any new product meet all technical and cost specifications in one run, modifications are usually required or complete redesign may be necessary. Hence the product may oscillate many times between design and testing (iteration) before an acceptable product design emerges. Close interaction with production and sales departments is usually necessary at this stage to ensure the product has a good chance of wide acceptability in terms of cost and quality. This is particularly important if there is already a potential competitor out there.

1.6.1.6   *Communication of Design*

The design engineer communicates through sketches and calculations as briefs for the draftsman who will develop detailed working drawings of component manufacture and assembly that will be used by the manufacturing facility. Samples of the product are usually tested regularly by the quality control department to ensure consistency in quality.

1.6.2   The Engineer's Factor of Safety (Factor of Ignorance)

No matter how detailed or thorough a design is or how capable the designer, the product can never be perfect primarily because there are always many unknowns and assumptions that have to be made in the various stages of the design process. Also, none of the personnel involved in the design and production process is perfect, and no one has succeeded in accurately predicting the potential effects of the service environment. In effect, the design engineer makes two major specifications: *tolerance, (or level of precision in manufacture)* giving acceptable limits of deviations from specifications; and *factor of safety (or factor of ignorance)* that mandates increments above the calculated design values to compensate for the unknowns. Variation beyond tolerance specifications may cause a product to malfunction or fail in service, and examples include dimensional or temperature variation limits. Specifications of tolerance and factors of safety are usually determined based on the critical nature of the component or product. For example, strength specifications of a critical component on an airplane, spacecraft, or a bridge support pillar may be increased by 100% or more to compensate for unknown variables and reduce the chances of potentially catastrophic failures in service. On the other hand, a much lower compensation would be required on a farm tractor component.

Inadequate factor of safety can lead to failure in service, with disastrous consequences. For example, in the design of a bridge, the traffic flow rate would have been determined in the process of gathering data by measuring traffic flow over a specific period of time. However, an accident could stop traffic on the entire length of the bridge and increase the load factor by several hundred percent compared with that imposed by moving traffic. Furthermore, there are many other unknowns such as the potential effect of corrosion, erosion, earth tremors etcetera on supports, hence it is not unusual to increase calculated values by a factor of three or more. However, it should be noted that tight tolerances or generous factors of safety can increase costs significantly and it is frequently a bone of contention between the design and production departments of manufacturing enterprises. It is not uncommon for manufacturers to deliberately lower safety factors and other recommendations by the design department, just to reduce costs. While they often get away with it, some serious disasters are believed to have been caused by this problem.

1.6.3   Performance Integrity in Engineering Design

The integrity of engineering design has increased exponentially in the past fifty years or so, facilitated by intensive and extensive research and development, and the emergence of the computer. An enormous database is now available on a wide range of inputs, from the characteristics of the natural environment, through the many artificial environments created by humankind, to the behavior of traditional and new materials under different environmental and loading conditions. While this development has greatly reduced the level of factors of safety required in design, it has not completely eliminated the potential for failure in service. There have been many disasters over time that show the inherent imperfection of engineering design. The British ship 'Titanic' was designed to be unsinkable but it hit an iceberg and cracked up on its maiden voyage from Southampton to New York in 1912. Before then no one knew that steel could lose up to 75% of its room temperature strength when submerged in freezing water. There have been many other monumental design failures in history in spite of near-zero failure tolerance applied in the design process. The U. S. Space Shuttle, Challenger disintegrated a few seconds into the flight in 1986 due to the failure of a low-cost O-ring seal used in the joint of one of the rocket boosters. This compromised the joint and allowed burning pressurized gases from within the booster rocket to escape and set the spacecraft on fire, killing all the astronauts on board.

Another Space Shuttle, Columbia disintegrated upon re-entry into the Earth's atmosphere in 2003, believed to have been caused by a small piece of foam insulation which detached from the fuel tank during launching and hit a wing of the spacecraft, damaging the silica plate thermal insulation of the wing. The spacecraft substructure could not withstand the very high temperatures on re-entry and disintegrated, killing all the astronauts on board. Three major nuclear accidents have occurred due to design failures and human errors in the last forty years, (Three Mile Island, U.S. in 1979; Chernobyl, Ukraine in 1986, and Fukushima Daichi, Japan in 2011). The Boeing 737 aircraft is the world's most successful airplane ever, introduced in 1968 and there are over ten thousand in service today, but two fatal accidents with almost five hundred fatalities within a few months of each other are believed to have been caused by a very simple design flaw in the software that controls a safety feature introduced in the latest redesign of the new generation series 737-700/800/900 in 2017.

1.6.4 Consequences of Engineering Design Failures

The effect of design failures can range from inconsequential to monumental: nearly two thousand lives of the world's richest were lost in the Titanic disaster, fourteen astronauts and civilians died in the two space disasters, and nearly five hundred lives were lost in the recent Boeing 737 crashes. Nearly twenty thousand people died or went missing as a result of the Daichi nuclear accident and, while very few lives were lost directly in the two others, the extensive and lethal radiation effects on life in the location areas are still being dealt with today. The good news is that each disaster is investigated comprehensively and designers learn a lot from the results which help to improve safety in the long term. This has led to the emergence of a specialization in engineering - *forensic engineering*. The Titanic disaster led to a comprehensive evaluation of metal behavior in every conceivable environment, and opened up a new subject in mechanical engineering - *fracture mechanics* - which has become a major tool in engineering design and greatly minimized failure of structures in service. The disaster also led to the development of a new range of alloys that can withstand extreme environmental conditions.

An O-ring is a very simple round ring made from an elastomer (rubber, polyurethane, silicone, etc.) used very commonly in sealing mechanical joints such as pipe connections to ensure fluid-tight seal between two sections, for example in garden water hoses. The O-ring seals used in a joint of the space shuttle Challenger's solid booster rocket were designed with a safety factor of three but they were susceptible to low temperatures and had lost significant strength during previous launches in cold weather. Forensic investigation showed that the near-freezing temperature on the day of the launch (around 2 degrees Celsius) was the last straw, a seal failed under the very high compressive stresses on lift-off and the escaping hot gases set the shuttle ablaze. Unfortunately this accident was complicated by human error: the commission that investigated the accident found that some engineers were aware of the potential issue, and had advised against launches in ambient air temperatures below 53 degrees Fahrenheit (12 degrees Celsius) but management decided to go ahead anyway.

The Columbia Space Shuttle disaster was caused by a piece of foam insulation which broke off from the Bipod ramp of the Space Shuttle's external tank at launch and struck the left wing of the orbiter, damaging its silica tile thermal insulation and exposing the wing's aluminum infrastructure to extreme temperatures. NASA engineers were aware of this incident but similar occurrences had happened in several earlier flights with no significant consequences, hence the impending disaster was not expected. On re-entry into the Earth's atmosphere, the orbiter's internal aluminum wing structure was exposed to extreme temperatures around 2,000°C causing the craft to disintegrate (aluminum melts at 660°C). Foam insulation is very light and is not considered to be a structural material that requires intensive design. However a broken piece of foam traveling at around Mach 2.5 becomes a powerful projectile. Furthermore, although silica tiles are very effective thermal insulators, they have low impact resistance. Forensic investigation revealed that the foam-insulated ramps fitted on the external fuel tank to reduce aerodynamic stresses were not even necessary and were removed from the tank in subsequent designs.

The angle of ascent of an aircraft is critical: if it is too high the engines will stall, if it is too low the plane will crash (the reverse is also true for descent), and pilots are well-trained to choose the appropriate values on take-off or descent, based on visual reading of instruments. However the Boeing 737-700/800/900 which came into service in 2017 features a new automated system designed to take over control from the pilot if necessary, based on angle of ascent/descent readings from two sensors mounted on either side of the plane. The monumental design error as determined by forensic investigation was apparently in the software which was designed to activate the automated system on the basis of the reading from either sensor. In both accidents, one sensor appeared to have malfunctioned, feeding erroneous data which activated the automated system, which in turn took over control from the pilots, turning down the plane's nose. A simple software design feature that requires action based on readings of both sensors coupled with the ability to easily deactivate the auto system if necessary could have prevented the accidents, and this is in fact the solution that the manufacturer has introduced, but the accidents led to the grounding of hundreds of the model around the world for many months. Furthermore, many pilots of that type of plane were not aware of this new feature and those who were only had a few hours of computer or simulator training. The manufacturer has also introduced statutory training programs specific to the aircraft model.

Engineering design has matured and the pace of advancement has been exponential in the last fifty years or so. This explains why, in spite of the very large increase in air travel, crashes of commercial airplanes are now very rare events compared with two or three decades ago. In fact, the odds of dying in an auto crash are nearly a hundred times higher than in an airplane crash.

Engineering is a human activity and therefore prone to both design and human errors. In the first two decades of the last century, machine failures accounted for around 80% of air crashes, but today, the statistic has reversed: about 80% of crashes are due to human errors (pilots, air traffic controllers, mechanics, etc). Many design products probably will never work the way they were designed to work. Often failures are slow and may be imperceptible for long periods until breaking point is reached (like fatigue failure in structures) but they may also happen instantaneously due to material flaws, design error, or unanticipated stress situations such as the impact of hurricane, floods, or earthquake, with disastrous consequences. However, engineers learn a lot from each accident which becomes very useful in reducing the chances of future re-occurrences. The bottom line is that perfection in engineering design is unachievable but near-perfection is possible.

2   Engineering Training

2.1   INTRODUCTION

The primary role of engineers is to design products and develop solutions to society's challenges. There are no templates for engineering education and many institutions tend to offer courses that prepare their products adequately for the needs of potential employers in their localities or country. Also, engineers will eventually be required to take on responsibilities outside of their traditional sphere, including leading teams, managing projects, or becoming chief executives of complex commercial endeavors, and therefore need to be familiar with basic management principles and social issues in addition to the core subjects. Many programs now feature courses in economics, principles of management, project design and planning, systems management, thereby giving their products a basic yet versatile foundation on which they can build in later years. Engineers may also end up as researchers, consultants, technical sales engineers, or in engineering education. In some developing countries, it is not unusual for fresh graduate engineers to be required to assume responsibilities that require experience and for which they are not adequately prepared, such as managing power plants or water works. For this reason, many institutions in the developing world place considerable emphasis on industrial attachment in the course of engineering education.

The structure and content of engineering programs vary widely between institutions and different specialties are often offered within the same department. However, there are commonalities, especially in the entrance requirements and core programs which focus on a strong foundation in mathematics, physics, chemistry, engineering mechanics, thermodynamics of solids and fluids, engineering materials; engineering drawing and design, with complementary laboratory experiments. Some institutions run the Sandwich Course System which requires students to spend one year in college and the next in industry until graduation. Others offer five-year master's degree programs with industrial attachment in the third year. These systems tend to produce competent engineers who blend easily with industry on graduation.

2.2 ADMISSION REQUIREMENTS

High school graduates with five or six good grades including mathematics, physics and chemistry may be admitted to first year basic science programs and have to pass mathematics, physics and chemistry (the equivalent of advanced level) before proceeding to engineering programs. Students who already have good grades at advanced level in these subjects, or mathematics, further mathematics and physics are admitted directly to the engineering program. However chemistry at advanced level is a requirement for some specializations such as chemical engineering, materials engineering, energy engineering. Food engineering and bioengineering require both chemistry and biology. Some programs also admit college and polytechnic diploma graduates whose qualifications are deemed to be equivalent to advanced level standards. Such students may already have qualifications in other relevant subjects such as engineering drawing and machine shop practice. Under the Course Unit System, diploma graduates can carry their credits to the degree program.

2.3 DURATION OF ENGINEERING PROGRAMS

The duration of programs also varies depending on institution and country. Most institutions run 3-year programs (four years in some institutions) for students with A-levels, and those with O-levels spend four or five years. In institutions that run the Course Unit System, duration can vary either way by 1-3 years depending on the ability of the student to acquire enough credits to graduate. Some engineering programs have a mandatory one-year industrial experience usually in the penultimate year to graduation, while others insist on at least two long-vacation industrial attachments before graduation. Most countries have accreditation bodies that set examinations primarily on the practical and professional aspects of engineering which a graduate engineer must pass, usually after a two-year industrial apprenticeship in order to become registered and competent to practice.

2.4 COURSE STRUCTURE

Most engineering programs have core courses (including studio, laboratory and workshop courses) which every student must pass before proceeding to a specialization. A typical basic engineering program includes:

- Engineering mathematics
- Engineering mechanics
- Mechanics of materials
- Engineering materials
- Mechanics and dynamics of fluids
- Thermodynamics
- Electrical engineering
- Computer science
- Engineering drawing/workshop practice/lab work
- Principles of engineering practice
- The engineer and society

Most   institutions also have auxiliary/general studies courses such as management principles, which the student must pass in order to graduate. The areas of engineering specialization vary widely between institutions and nomenclatures differ, so are the combinations of specialties or sub-specialties. For example, some institutions combine electrical and computer engineering while others offer electrical and electronic engineering, civil and environmental engineering, electronics and communications engineering, chemical and metallurgical engineering, materials science and engineering, mechanical and agricultural engineering, mining and minerals engineering, etc. Departments which offer combined courses usually have core courses common to the related specialties after which a student spends the last year or two on courses relevant to a chosen area. Also, some institutions offer very specialized engineering programs such as aerospace, biomedical, architectural, structural, geotechnical, water, transport, energy, agricultural, food, petroleum, robotics, mechatronics, sports, nuclear, nanomaterials, biomaterials, etc. Most of them are sub-specialties of the main types or hybrids of several and will be discussed in Chapter 4.

2.5   CAREER OPPORTUNITIES IN ENGINEERING

Engineering is the prime propeller of economic, industrial and human development, and it is difficult to imagine an aspect of human well being that does not depend on engineering: building, manufacturing, machinery, energy, agriculture, water, appliances, consumer products, transportation, communication, health services, education, environmental health, entertainment, and many more. The engineer is often compared with the medical doctor in terms of responsibility to society. Indeed both professions are critical to the well-being of society but the responsibilities are not comparable, hence the common saying *"the doctor buries his mistake, the engineer is buried by his mistake"*. While a doctor's mistake may result in one fatality, the engineer's mistake could result in a large loss of lives: collapsed levees, platforms, bridges, buildings; road, train, aircraft and marine accidents; chemical disasters, and any of these could bury the career of an engineer. While a doctor's mistake may be missed, forensic investigation of a major disaster will likely identify causes and apportion blames. Engineers are known to have taken their own lives as a result of serious disasters blamed on faulty design. It is vital therefore that engineers are conscious of the potentially enormous impact on society of their successes as well as their failures.

The challenges and opportunities of engineering, and the central role of engineers in scientific and technological innovations have become critical to the global economy and well-being of society, and this enormous influence is creating an increasing demand for engineers which will inevitably grow exponentially, considering the growing global economy and population, and increasing diversity of societal needs in virtually all spheres of life. Also, engineers and scientists will be the primary movers of future technological developments and innovations on virtually every front of human endeavor. Globalization depends critically on transportation and communication technologies, creating new opportunities for engineers. Furthermore, most engineers will eventually end up in senior management positions with responsibilities for managing large complex projects and many employees, a development which helps to make the profession a tremendously rewarding endeavor. Many of the leading industrial, engineering and manufacturing enterprises of today (Rolls Royce, General Electric, IBM, Microsoft, Apple, Hewlett Packard, Compaq, Intel, Google, Oracle, Yahoo, Xerox, Dyson, etc.) were founded by engineers as very small businesses decades ago, and entrepreneurial opportunities for engineers are growing exponentially. Details of potential career opportunities are discussed under each specialty but every engineer also has major opportunities in education, research and development, and many are involved in consulting and technical marketing/sales.

THE ENGINEERING FAMILY

MILITARY
Civil
Mechanical
Electrical
Logistics
Strategic Support

CIVIL
Construction   Foundation
Structural   Building
Highway   Environmental
Transport   Urban
Water

MECHANICAL
Manufacturing   Marine
Industrial   Aeronautical
Automobile   Power
Agricultural   Petroleum
Mining   Mechatronics

MATERIALS
Metals   Nanomaterials
Polymers   Minerals
Ceramics   Corrosion
Composites   Forensic

CHEMICAL
Refinery
Petrochemicals
Process
Pharmaceutical
Food

ELECTRICAL
Power
Electronics
Control

COMPUTER
Hardware
Software
Systems
Cyber-Security

TELECOM
Hardware
Network
Broadcasting

NEW GENERATION
(Hybrids of several traditional branches)
Nuclear   Energy
Robotics   Biomedical
Geotechnical   Systems
Architectural   Sports
Aerospace   Nanotronic
Artificial Intel   Forensic

3 Engineering and Society

Engineering needs to focus on changing the world for better,
rather than on technical ingenuity for its own sake

3.1. INTRODUCTION

Technology embodies skill sets, techniques, methods and processes required for the production of resources that are critical for human development, and engineers are the prime movers. They design products and production processes; they design and build machinery, structures and buildings; they manage production; they drive innovation. The prime role of technology dates back to the beginning of human culture when society always sought to transform the natural resources and materials in the world around them to create new technologies that complement human effort, and throughout history, technology and society have co-evolved. Today, the symbiosis is such that societies shape technologies, and technologies also shape societies. Technology innovations are driven by societal needs and wants, while new and emerging technologies create new needs and wants. Human development has benefited tremendously from technology but the negative impacts and unanticipated consequences on society and the environment are becoming increasingly apparent as technology becomes more pervasive or powerful.

The interdependence of technology and society is very broad, touching almost every technology and every aspect of human development and raising a variety of societal issues and concerns.

All technology innovations come with consequences and trade-offs many of which are having increasingly negative impact on society. It is vital therefore that the creators and prime movers of technology innovation should also be aware of the potentially negative impacts on society, especially because many engineers will rise to management positions and have to take technological decisions which require making choices between competing priorities. For example, the traditional approach to design needs to change. Currently, engineers design new products to meet specific needs. The evolving, environment-friendly approach is to consider the entire life cycle of the product: sources of raw materials, industrial processes and energy requirements involved in production, process and product efficiency issues, service integrity, and end-of-life disposal. Engineers need to be aware of the environmental benefits of producing durable products using fewer resources, minimizing potential environmental pollution, and the benefits of reusing and recycling. Two technologies - energy and information technology - permeate virtually all aspects of corporate, industrial, social and personal activities and, while many other technologies have potentially negative impact on society, these two are of the greatest concern and are discussed in some depth in the next section.

3.2   ENERGY TECHNOLOGY AND THE HUMAN ENVIRONMENT

Coal provided the energy for the Industrial Revolution starting from the sixteenth century AD. It supplied the energy for mass production of steel, non-ferrous metals, cement, steam engines, steam locomotives; coal gas powered the early automobiles, and electric power generators. Development of the oil and gas industries from the beginning of the last century provided energy for the modern transportation systems. Electricity generated mostly from fossil fuels (coal, natural gas, oil) has become indispensable in virtually every aspect of human development. However, the negative impacts of these developments on the environment (contribution to global warming with its negative effects on global climate and human health issues) have accentuated in the last seven decades, synonymous with global economic and industrial growth, to the extent that the world is now dealing with the severe consequences: extreme weather, desertification, respiratory and heart diseases, ecological issues.

Rapid economic and industrial growth in most countries of the world is causing unprecedented demand for energy and most depend on locally available or easily procurable resources. For this reason, fossil fuels currently provide over 80% of global requirements of primary energy, and this near total dependence is unlikely to change for many years to come despite global efforts for several reasons: alternatives (hydro, solar, wind, geothermal) also have issues which have stunted growth and scope of applications, limiting combined contribution to primary energy supply to less than 20%. Furthermore, the fossil energy industry sustains the economies of many countries across all continents and economic classifications of the world, in terms of job creation and foreign exchange earnings (Afonja, 2020). Many critical products: steel, cement, chemicals, fertilizers depend on fossil fuels as energy source or precursors. Petrochemicals derived from fossil fuels have become indispensable to modern living: plastics, high-technology materials, pharmaceutical products, construction materials, agricultural fertilizers, consumer products, automobiles, airplanes, appliances, clothing, furnishing. Coal, the most polluting fossil fuel is the predominant energy resource in many emerging countries which also account for the fastest growth rate in primary energy demand and associated environmental pollution, as well as in China, the world's highest source of energy-related pollution. Nevertheless, the current global drive to clean up energy will inevitably create new opportunities, and new challenges for up and coming engineers.

The environment has benefited in many ways from technology innovations, but the negative fallouts can be potentially more devastating especially because of the time lag between cause and effects. The negative effects of much of the pollution being generated today may not become evident for decades, could persist for hundreds of years, and could resonate in global locations far removed from the source. Most engineers will at some point in their careers find themselves in positions where they can influence policies and technologies that protect the environment, such as process energy and transportation fuel efficiency, material and energy recycling, green energy development, carbon capture, recycling and sequestration.

3.3. INFORMATION TECHNOLOGY AND SOCIETY

Rapid advances in information and communication technologies from the later part of the twentieth century are creating a major Information Age revolution, with the possibility to digitize almost any information, thereby facilitating easy management, storage, access through a wide array of devices in various formats, and transmission routes. The impact on individuals, organizations and entire societies has been seismic and profound in many ways. Information technology is a basic prerequisite for critical human development challenges such as economic development, industry, commerce, healthcare, education, politics, social interaction. Socio-economic globalization, growing multi-continental business enterprises, e-commerce, the emerging norms of home-based work would not have been possible without the Internet and World Wide Web. More than half of the world population spend significant time daily doing business, accessing entertainment, or socializing on line, leaving troves of digital footprints and virtually oblivious of the potential consequences. New methods of communication: cell phones, emails, social networking, virtual meetings of people residing in different parts of the world, have revolutionized human interaction worldwide. Industry, health, education, commerce, transportation are now all so dependent on technology that any disruption such as cyber-crime or systems collapse can shut down thousands of enterprises across many countries of the world.

Three information technology innovations in the last four decades or so have galvanized technology and society to such an extent that virtually every aspect of human development is now dominated by technology: invention and proliferation of personal computers in the 1970s, invention of the Internet starting from the early 1960s until maturity in 1989, and proliferation of smart phones from the mid-1990s. The personal computer made it possible to digitalize, process and store information and data; the Internet/World Wide Web connected the world; and the smart phone equipped almost everyone in society with the wherewithal to communicate, do business and socialize on line. Currently, almost a third of the world's 6.8 billion population use the Internet regularly. Smart systems and devices (Internet of Things, IoT) dominate almost every aspect of the lives of modern society, to the extent that they are indispensable in industry, transportation, healthcare, education, agriculture, social life. An estimated ten billion IoT devices are connected to the Internet in 2021, regulating temperatures in buildings, controlling traffic, monitoring movements, protecting vital installations, controlling weather monitoring and military attack drones.

3.4. ETHICAL AND SOCIAL ISSUES WITH TECHNOLOGY INNOVATIONS

The impacts of technological advances on virtually all aspects of human development have been largely positive and beneficial but the negative fallouts are also increasingly serious. Environmental pollution from energy use in industrial, transportation and housing sectors is exacerbating global warming, the negative consequences of which may resonate across continents and remain potent for centuries. Storms are becoming more frequent, much more powerful, more destructive and enduring because they derive their energy from the warming oceans which occupy three quarters of global land and serve as heat reservoirs. They originate mostly from tropical waters but make devastating landfalls thousands of kilometers from source. Heat waves and associated deaths are becoming increasingly common and intense because increased pollution mostly from energy use is enhancing the heat-trapping capacity of the natural greenhouse gases which control the Earth's temperature. Pollution also accounts for about 3,5 million deaths from lung and heart diseases annually, and women and children are the most vulnerable.

The information revolution of the last three decades or so has had very positive impact on society in many ways: it has changed dramatically the way people interact and communicate with each other daily, with a very wide range of new methods - people can communicate through emails, video conferencing, social networking, or facetime a person anytime, almost anywhere in the world. Information technologies have greatly enhanced the process of learning by facilitating communication, interactive and collaborative learning beyond the classroom, with information available on a 24-hour basis. It has greatly improved healthcare in many ways, notably through the proliferation of diagnostic instrumentation, digital implants, robotic surgery, telehealth systems, virtual personal well being monitoring facilities, etc. Industry, commerce, transportation, home management depend critically on Internet of Things (IoT), and artificial intelligence (AI) is playing an increasingly prominent role in virtually all aspects of our daily lives. However, the current prime role of information technology in our daily lives has also raised many social, technological, and ethical issues. Society affects what information and knowledge is available, where, when and how it is used, but the ready availability of information could impact society in potentially negative, profound and unanticipated ways. Society is fast losing privacy, control over who accesses the vast information resources and how they are used. Invasion of privacy is emerging as a major issue: corporate and governmental power to surveil users of information technology, harvest and use sensitive information is of increasing concern; Internet users are followed through cookies by every website they visit; cyber-interference in local politics and the affairs of other countries is now common; digital information has emerged as a potent tool of surveillance, behavioral manipulation, radicalization, and digital addiction.

Online security is emerging as a major problem. Stealing of sensitive personal information, emergence of the Dark Net haven for criminals, cyber-disinformation, cyber-bullying are all by-products of technology innovations. Apart from the fact that almost everyone is now a potential victim of cyber-crime, the amount of personal and sensitive information collected legally and illegally daily through social media, surveillance equipment, artificial intelligence, tracking of Internet web activity, is astounding. Highly skilled cyber-criminals have emerged all over the world and cyber-security has emerged as one of the most challenging issues of the modern world. Independent and state-sponsored cyber-attacks are disrupting major enterprises and utilities in other nation countries and meddling in their political affairs. The potential for intensified, inter-state cyber-warfare is becoming a reality and businesses and critical utilities are becoming prime targets.

Technology innovations are also raising other ethical issues and concerns. They are changing lifestyles all over the world, with increasing automation and sedentary lifestyle on many fronts: from remote switching the television on and off, to spending hours on digital viewing and social media. IoT devices and robots are becoming increasingly prominent in modern homes. Overuse and compulsive use of digital technologies have potentially nega-tive physical and psychological health effects. Texting while driv-ing is now a common cause of fatal road accidents. Most everyday digital technologies are sedentary and extended use promotes sedentary lifestyles. There is ample research evidence that reduced physical activity and increased digital screen addiction are promoting many health, social and psychological issues such as obesity, cardiovascular disease type 2 diabetes, depression, anxiety, feeling of isolation, suicidal behavior, sleep problems, poor posture, eye strain, destruction of social fabric, diminishing social skills, etc. The mind space of the youngest generation is particularly vulnerable: they spend a lot more time on cell phones, social media and video games; they are impressionable and could succumb relatively easily to negative influences. Recent studies involving children and young adults found that, on average, they spend five to seven hours a day glued to one electronic screen or another but only have one to two hours of exercise a week. Also, people with higher social media use were more than three times as likely to feel socially isolated than those who did not use social media as often (Primack *et. al,* 2017).

The human mind plays a critical role in our daily survival: it perceives and evaluates the environment and takes appropriate actions to achieve a desire. This natural instinct (the ability to perceive, hear, think, and act appropriately) is being replicated in machines and constitutes a new field of artificial intelligence (AI). The home thermostat measures temperature of the environment, compares with desired setting, adjusts the air conditioning system appropriately, and can be controlled remotely from anywhere in the world that has Internet access. Robots known as humanoids are becoming increasingly prominent on industrial assembly lines, performing critical operations in health systems, managing customer care services, featuring prominently in military warfare, and taking over many of the traditional roles of human beings in transportation.

Robots are performing dangerous jobs such as defusing bombs and attacking enemy targets, or taking over boring and repetitive jobs, for example on industrial assembly lines. Driverless shuttle trains and buses are common features of airports, amusement parks, etc. Already, over half of the traditional human jobs are now being performed by connected devices and robots, and the ultimate goal of AI innovations is to completely replace human efforts so that people get paid a basic wage for no work. This makes economic sense because robots are a lot more reliable and productive than human effort. However, technology is great when well blended with human effort but overuse or misuse could have serious consequences. Human intelligence is not infallible and this replicates in artificial intelligence which still needs to be designed, built and programmed by humans. Furthermore, just like humans, it is impossible to program machines for every possible eventuality. This explains the many recent accidents involving autonomous automobiles auto-piloted entirely by artificial intelligence while drivers engaged in socialization. This is a clear case of misuse because autopilots still require human supervision. The goal of entirely replacing human effort with artificial intelligence also raises potentially serious ethical questions on the prospects of rising unemployment and the consequences of breeding an idle society.

3.5. THE ENGINEER AND SOCIETY

Most students choose to study engineering not because they know much about what engineering involves but because they are good at STEM subjects, they see it as a good, well-paying career, or have skills that they believe will be well suited to a career in engineering. The name 'engineer' defines expertise in designing, building, maintaining, and working with engines and other structures; and the scope has recently extended beyond hardware to include software and virtual intelligence. The discipline requires ingenuity, creativity, imagination, innovation, and perseverance. However, the products of engineering are designed to fill the needs of society and, inevitably, engineers will, at different points in a career profile, have to interact and work with people. This requires communication, negotiation, judgement, and management skills; and these characteristics will become increasingly vital as they rise to management positions.

It also requires that engineers are aware of, and develop skills to manage the potentially negative impacts of their creations on society: safety, environmental degradation, ethical issues. The strong growth in global population, urbanization, and increasing demand for food, water, shelter, education, healthcare, transportation, consumer products, all provide the engineer with new and formidable challenges, and engineering education has to be dynamic to be able to respond appropriately. Engineers need to be aware of, and develop skills to manage the potentially negative impacts of their creations on society: safety, environmental degradation, ethical issues. No matter how thorough an engineering design is, there will always be failures, some catastrophic because no design can be perfect, and the first instinct of the producer is to protect personal or corporate integrity by concealing details that could reveal the cause. Professional ethics demand absolute honesty and cooperation with forensic investigation, the result of which could inspire better design and innovations that save lives. Also, quite often the engineer is under pressure from management to make design compromises that cut production costs. The result could be a catastrophic component failure, (for example in an aircraft, high-speed train, bridge, building, chemical plant) causing huge loss of lives. Integrity is central to the engineer's code of conduct.

The Engineer's Code of Ethics

Engineering has a direct impact on the quality of life of all people and engineers must perform under a standard of professional behavior that requires adherence to the highest principles of ethical conduct

Professional services provided by engineers require honesty, impartiality, fairness and equity, and must be dedicated to the protection of the public health, safety, and welfare.

Engineers must perform services only in areas of their competence, exhibit the highest standards of honesty and integrity, conduct themselves honorably, responsibly, ethically, and lawfully so as to enhance the honor, reputation, and usefulness of the profession

(Extracted from the Code of Ethics of the American Society of Professional Engineers)

4   Specialties in Engineering

4.1   INTRODUCTION

Prior to the late 1800s, there were four broad specializations in engineering: military, civil, mechanical and metallurgical. Military engineering, a combination of many specialties, is the oldest of them all, evolving from Early history wars. It involves designing, building and maintaining military equipment and infrastructure (weapons, buildings, roads, bridges, transport, workshops), communications, military logistics, and all other operations that shape the military environment. Civil engineering evolved from military engineering when the same techniques and strategies began to be applied to civil structures. Although the Early people had been smelting metals from ores for thousands of years and the Romans used bronze extensively in armory, it was the availability of iron and steel on a mass scale (metallurgical engineering) made possible by the emergence of commercial coal mines and coking technology, that stimulated widespread design and construction of machines, mechanical equipment and structures, the humble beginning of mechanical engineering. The invention of electricity and growth of the chemicals industry towards the end of the nineteenth century AD gave birth to two more areas: electrical and chemical engineering. Since then, metallurgical engineering has largely transformed into materials engineering because of the widening scope of use of non-metals, and many new areas have emerged, most of them sub-specializations of the core areas, or fusion of two or more traditional specialties. Examples include structural, agricultural, mining, automobile, traffic, aeronautical, marine, petrochemical, telecommunications, aerospace, robotics, rocket science and engineering, etc. Emergence of polymers (plastics), commercial nuclear energy, transistors and computers in the 1950s also added several more specialization areas.

Advances in engineering and medical sciences have identified the very close relationship between the human body and engineering and many basic engineering phenomena also feature in the human body. For example, the heart is basically a mechanical fluid pump; the lung operates on similar principles as the mechanical air box bellows; the same laws that govern flow of fluids in a pipe are applicable to blood flow in arteries and veins; and most joints in the human body (ball and socket, hinge), obey the same laws of mechanics as joints in the sub-structure of an automobile. Furthermore, it is increasingly possible to replace natural human parts with artificial electro-mechanical complements or substitutes, (dentures, heart valves, hip joints, arms, legs etc.). The work of kidneys can now be performed by mechanical dialysis machines; worn hip joints are now being relined or replaced in much the same way as the automobile brake disk/drum; complete mechanical hearts are under development; and humanoid/robotics technology is a first step in replicating the human brain. Corrective, reconstructive and cosmetic medical surgeries now use a very wide range of implants made from foreign materials in the highly corrosive inner body environment. These developments have stimulated the emergence of other specializations, notably biomedical engineering and bio/nano materials engineering.

Artificial intelligence is playing an increasingly important role in industry, transportation, home management, military operations, and giving birth to new specializations such as robotics or mechatronics engineering. The most common specializations in engineering are discussed in the following sections. However, it should be noted that the classifications are not unique. For example, whereas all engineering is deeply rooted in science, many programs still feature science in nomenclature for emphasis, such as materials science and engineering or computer science and engineering.

4.2   CIVIL ENGINEERING

As mentioned earlier, construction (the core of modern civil engineering) has been practiced from the Early times. Caves were dug in mountains and huts were constructed from wood, stones, mud or clay, with thatched roofs made from raffia palm. It is interesting to note that the rudiments of structural and composite materials engineering were in common practice: mud walls strengthened with wood stakes and stones, straw embedded in clay, and roofing strengthened with woven mats. Pyramids, city walls and other complex civil infrastructures were prominent features of ancient cities five thousand years ago. Wars have always been part of life from Early times and rudiments of civil and mechanical engineering featured prominently in logistic planning and attacks (roads, bridges, transport, weaponry). Over time, civil engineering has evolved as the broadest of engineering disciplines, meeting challenges of building and transportation infrastructure, urban development, utility provision, environmental pollution, etc. Civil engineering features in a wide spectrum of societal life: planning, design, construction, maintenance and management of facilities essential to modern life, such as buildings, roads, bridges, tunnels, airports, transportation systems, hydro-dams, water treatment and distribution, industrial concrete structures and foundations, drainage systems, sewage and pollution management, environmental control and remediation. Modern civil engineering currently has six main sub-specialties (Figure 4.1). All of them still feature to some extent in a typical civil engineering curriculum, hence it is relatively easy for a graduate civil engineer to specialize in any of the areas.

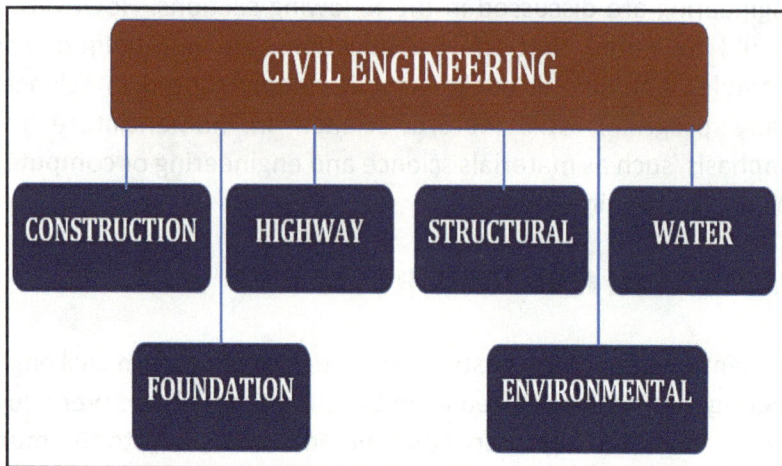

Figure 4.1.   Sub-specialties in Civil Engineering.

In the earlier times, civil engineering was primarily about construction of buildings, roads, dams and most decisions were based largely on intuition and experience. However, with the development of basic engineering concepts and theories, the design of foundations and structures is now based on systematic processes involving extensive theoretical calculations to determine appropriate specifications for structures required for buildings, roads, airport runways, foundations, which has led to the relatively new sub-discipline of structural engineering. Furthermore the increasing demands of human development have given prominence to the provision of water and a conducive environment, thus further expanding the scope of civil engineering to include relatively new sub-specialties: water and environmental engineering. Prior to the emergence of quantity survey as a profession, the civil engineer also had the responsibility for managing all costs relating to building and civil engineering projects: the initial cost estimation and cost control to ensure that target specifications were achieved at optimal cost. However, most civil engineering programs still impart basic knowledge in cost estimation and control.

4.2.1 Construction Engineering

Construction is the 'heart and soul' of civil engineering, and most other specialties in engineering involve some civil engineering construction: buildings, foundations, transmission towers, etc. Apart from competence in basic engineering, the construction engineer must be able to read design drawings prepared by the architect, structural, mechanical and electrical engineers, and faithfully create structures and services in accordance with specifications. A good working knowledge of the major materials of construction and stability in different environmental situations is vital: cement, concrete, reinforced concrete, mortar, wood, steel, ceramics, polymers, glasses, composites, paints and coatings, etc. Training includes choice of appropriate concrete mixes for different applications in construction (foundations, floors, walls, columns, etc.), setting times, drainage and sewage requirements, roofing materials, ventilation, plumbing, lighting, etc.

4.2.2 Structural Engineering

Structural engineering is a major branch of civil engineering which deals with the structural integrity of buildings and other structures: foundations, roads, bridges, water impoundment dams, levees, towers, etc. The profession dates back to the Early history around 3000 BC when huge pyramids were constructed in Egypt. Pyramids were popular because of the inherent structural stability and infinite scope for scale-up, and stones were used in the construction because of the high compressive strength. The high structural integrity of pyramids explains why some are still standing today. Structural design in these times was based largely on intuition and experience, and a theoretical basis only started to evolve during the Industrial Revolution with the proliferation of the use of concrete and reinforcement steel in commercial quantities, and building of large structures. Although every civil engineer has basic training in structures, structural engineering has evolved as a distinct and vital sub-specialty in view of the increasing complexity of civil structures.

The work of the structural engineer cuts across all other sub-specialties in civil engineering and, indeed other engineering specialties and architecture. It involves assessment of soil integrity and design of appropriate foundation; calculation of loads and stresses (static, dynamic, tensile, compression, bending, complex) that a construction component or structure may be required to withstand in service, taking into account environmental conditions, and the determination of appropriate dimensions and reinforcement. Each major section of a construction requires specific concrete mix ratios and some steel reinforcement, and reinforced trusses, pillars, columns may be required. The structural engineer must determine the appropriate concrete mixes for specific areas of a structure, specifications for sectional dimensions and arrangement of reinforcement steel, support pillar structures, etc. High rise buildings, bridges, roads that pass through marshland, structures immersed in saline water, buildings and structures in areas prone to extreme weather, flooding or earthquakes, all require special structural design and reinforcement. The design calculations and specifications are presented to the technician/draftsman who prepares detailed working drawings that can be interpreted and executed by the construction engineer. Also, the structural engineer is normally required to oversee construction projects to ensure that design specifications are adhered to.

Structural engineering involves complex technical analyses, extensive and intensive mathematical calculations. Strong competence in structural mechanics, mechanics of materials, and engineering mathematics is a primary prerequisite. Structural engineers need a working knowledge of soil mechanics, and a good working knowledge of materials-environment interaction is also essential: for example, concrete structures that will be immersed in corrosive saline sea water or freezing environment (like bridge supports), or structures in areas prone to tornadoes need special design in terms of appropriate dimensions, types and quantities of steel reinforcement, and protection specifications. The arrival of computers and appropriate software packages has greatly enhanced the work of the structural engineer, and computing skills are vital.

4.2.3 Highway Engineering

Prior to the development of the automobile industry in the early part of the twentieth century, there were only foot pathways but bridges still had to be constructed across rivers, and this was done by intuition, using wood, stone and soil. The advent of the automobile accentuated the need to construct durable roads, the humble beginning of highway engineering. Over time, road networks have become very complicated and sophisticated, bringing to prominence the sub-specialty. The work of the highway engineer is to design roads, from township roads to complex highway networks, complete with bridges, interchanges, overpasses and underpasses, appropriate road signs, efficient and safe traffic flow, traffic lights, walkways, rest areas, emergency services, etc. The highway engineer is either also a structural engineer, or must work closely with one to determine the static and dynamic loads that roads and bridges will carry and the types of structures and sub-structures required. A good working knowledge of soil mechanics is also mandatory to provide data for the appropriate structural design. The highway engineer must have expertise in paving materials technology (concrete, asphalt, etc.). Familiarity with environmental engineering, in particular, drainage and landscape design is also essential. The work of the highway engineer is becoming increasingly complex in view of the exponential growth in traffic and heavy trucking, and increasing constraints on land availability. Multi-level road networks and complex intersections are now common in many developed countries. Highway engineers also design and oversee the construction of airport runways, rail tracks, etc.

4.2.4   Water Engineering

Water is indispensable to life and provision of adequate supply is a priority in most human development initiatives. Municipalities usually obtain processed water from impounded rivers and streams, or boreholes. Saline ocean waters are also processed in some countries. The primary work of the water engineer is to determine suitable rivers/streams for impoundment, and, working with the structural engineer, determine a suitable location and specifications for the concrete impoundment structure. The water engineer designs the scheme layout, including the purification plant, pipe network, pump specifications, and makes appropriate provision for dealing with overflows. A good working knowledge of water purification technologies which vary with the initial physico-chemical properties of the impounded water, and competence in the design of appropriate flowcharts are also essential. Rivers are often impounded for other purposes: hydropower generation or recreation, fish ponds, swimming pools, barriers to control ocean surges and flooding (levees), etc., all of which fall within the purview of the water engineer. Communities also obtain water supply from sub-surface sources through boreholes. The water engineer needs to know how to interpret geological and geophysical data in order to appropriately site such water schemes and should have the competence to design and commission boreholes. Some communities also depend on processed saline (ocean) water which requires specialized equipment and processes, all within the purview of the water engineer.

4.2.5   Foundation Engineering.

Virtually every structure requires a foundation: buildings, roads, bridges, impoundment structures, communication towers, heavy machinery, etc. The foundation engineer is in fact a specialized structural engineer with competence in the design of appropriate foundations for different civil or mechanical structures. Also, in spite of comprehensive structural design, foundation integrity of buildings, roads, bridges, etcetera can deteriorate over time, caused by soil subsidence, erosion, earth tremors, or material failure due to oxidation, corrosion, creep, fatigue, etc. The foundation engineer must have the competence to carry out inspections and determine the integrity of structures in service, detect problems, and recommend feasible solutions. The foundation engineer must also have a good working knowledge of soil mechanics, structural mechanics, stress analysis, and materials and concrete technologies.

4.2.6 Environmental Engineering

The integrity of the environment has become a major issue over the last few decades, not just because of pollution or degradation but also because a conducive environment is becoming increasingly desirable. Provision of a beautiful environment has become a priority in the design of new housing estates and in many cities, including design of aesthetic, functional and human-friendly layouts; provision and location of appropriate road networks, amenities, recreation facilities, walkways, beautiful landscapes; drainage and flood control; sewage and refuse disposal; waste recycling; design, deployment, operation and maintenance of air pollution control systems; and many more. The increasing global focus on environmental pollution and the increasingly extensive need for analysis and monitoring of sub-micron pollution are also creating new challenges and opportunities for the environmental engineer. Most major new civil and industrial projects require comprehensive environmental impact assessment before approval. Environmental engineering is an evolving sub-specialty of civil engineering and requires competence in systems design, planning, analysis and deployment. However, inter-disciplinary skills are almost always required to resolve most environmental issues and graduates in the sciences, architecture and indeed any engineering specialty can easily retrain as environmental engineers.

4.2.7   Career Opportunities in Civil Engineering

Civil engineering is the oldest and broadest of the engineering disciplines, and one of the three most versatile specialties in engineering because of the way it features in every sphere of human development. Construction, structural, highway and water engineers are particularly versatile because their products sustain human development and feature prominently in everyday life: residential, commercial and industrial buildings, roads and bridges, water supply, recreation areas, etc. A civil engineer works with a wide spectrum of individuals in both the public and private sectors to meet today's challenges of pollution, infrastructure rehabilitation, traffic congestion, floods, earthquakes, and urban development. Civil engineers plan, design, construct, maintain, manage, and operate facilities essential to modern, civilized human life. Projects requiring civil engineering expertise vary widely in nature, size, and scope, and include bridges, tunnels, transportation systems, airports, storm water drainage systems, dams, buildings, foundations, water treatment and distribution, wastewater management, hazardous waste treatment, environmental remediation, environmental protection, and air pollution control.

With a good degree, a graduate of civil engineering can quickly specialize in any of the sub-divisions, either through on-the-job training or a graduate program. Civil engineers work in public and private enterprises, or as entrepreneurs/contractors. Structural engineers are particularly versatile because their expertise is required across many engineering areas, such as conduit structures, oil and gas rigs and pipeline networks, aerial masts in telecommunication, etc. However, the sub-specialty requires a strong flare for mathematics, engineering mechanics, mechanics of materials, materials of construction and computing.

The construction industry is one of the largest industries in the world, based on either employment or expenditure. A responsible position in construction management requires the ability to apply principles from business, mathematics, science, and engineering to a very wide variety of construction projects in terms of nature, type, and scope. Construction managers plan, construct, maintain, and manage facilities essential to modern, civilized human life. Projects requiring construction management expertise include buildings, bridges, tunnels, transportation systems, and facilities utilized in various specialized industrial processes. Most modern civil engineering curricula are broad-based and provide students with a strong foundation to facilitate on-the-job specialization or through graduate courses.

4.3 MECHANICAL ENGINEERING

The desire to complement human effort has stimulated the evolution of mechanical engineering from the Early times: working wood and stones to produce hunting and farming implements, crude rope lifting devices and wood cantilevers to hoist loads that were beyond human physical effort, waterwheels and wind mills to grind corn, etc. Conflicts and wars from Early times also led to the development of mechanical tools, animal chariots, and weapons. However, mechanical engineering really evolved with the advent of the coal industry which powered the Industrial Revolution. Although coal collected from outcrops had been used from Neolithic times, development of large-scale mining of underground coal deposits which began in the seventeenth century AD provided the impetus for the development of drilling, excavation and evacuation equipment, and pumps to reclaim water from flooded mines. Availability of coal in commercial quantities in turn set off a chain of events which stimulated the development of steam engines, rail transport; textile mills, agricultural machinery and implements; steel, cement and automobile industries; and a firm establishment of mechanical engineering as a profession, with several sub-specialties (Figure 4.2).

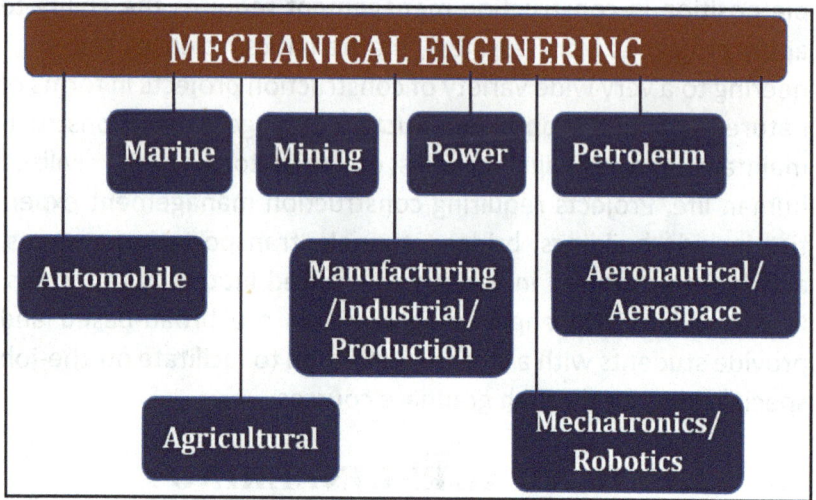

Figure 4.2. Sub-specialties in Mechanical Engineering.

The work of the mechanical engineer centers on the design, manufacture, operation and maintenance of machinery and associated equipment that produce goods: home goods, automobiles, airplanes, space craft, internal combustion engines, steam boilers, steam/gas/wind, turbines, civil construction machinery, mining and agricultural machinery, food and pharmaceutical processing machinery, oil and gas drilling and refining equipment, pipeline and rail transportation equipment, mechanical structures, pumps, military hardware, etc.

Mechanical engineers also design production processes such as casting, forging, machining, welding, rolling, drawing, briquetting, pelletizing, plating etc. Mechanical engineers are extremely versatile and can be found in a large variety of private and public sector organizations. They may be involved in product design and development, manufacturing, equipment maintenance, project management, oil and gas drilling, power generation, and many other operations. They apply a deep working knowledge of technical fundamentals in areas related to mechanical, electromechanical, and thermal systems to address needs of society. They also develop innovative technologies, find solutions to engineering problems, and lead in the conception, design and implementation of new products and production processes, services, and systems. There is hardly any consumer product (household goods, automobiles, pharmaceuticals etc.) that has not passed through several of these processes, or does not require the input of a mechanical engineer at some stage. Mechanical engineers design and oversee the construction, installation, commissioning, and operation of process plants, oil rigs, solar, hydro, wind, and geothermal power plants, water, oil and gas pipeline networks. They are also closely involved in around 80% of electric power production: fuel combustion, boiler, steam/gas/water/wind turbine design and operation, as well as the production aspects of oil and gas. Mechanical engineers require a solid foundation in core subjects: solid and fluid mechanics, thermodynamics, machine statics and dynamics, control systems, materials, applied mathematics, and engineering design. Over time, several sub-specialties of mechanical engineering have evolved and some of the most common are discussed briefly below. It should be noted that many curricula offer a foundation course after which a student can specialize in one of many areas: production, industrial, mining, automotive, agricultural, petroleum, power, aeronautical, mechatronic, robotic, railway engineering, and several more.

4.3.1 Manufacturing Engineering

Every product passes through many stages of production before emerging as a finished good and they all involve the use of machinery. Manufacturing engineering involves design, operation and maintenance of manufacturing flow lines and appropriate machinery: lathes, milling machines, grinding machines, presses, rolling mills, drawing benches, welding bays, forging presses, briquetting and pelletizing machinery, casting bays, plating sections, material handling equipment, drilling and excavation equipment, agro-machinery, unit operations equipment for the chemical industry, etc. Manufacturing is becoming increasingly mechanized and automated, thus increasing the demand for a wide range of automated machines.

4.3.2   Mechatronics/Robotics Engineering

Mechatronics engineering is an interdisciplinary branch of engineering that focuses on integration of mechanical, electronic, electrical engineering and control engineering systems (Figure 4.3). Mechatronic/Robotic engineers use a combination of mechanical, electrical, computer and computer skills to design and operate smart technologies, such as robots, automated guided systems and computer-integrated manufacturing equipment.

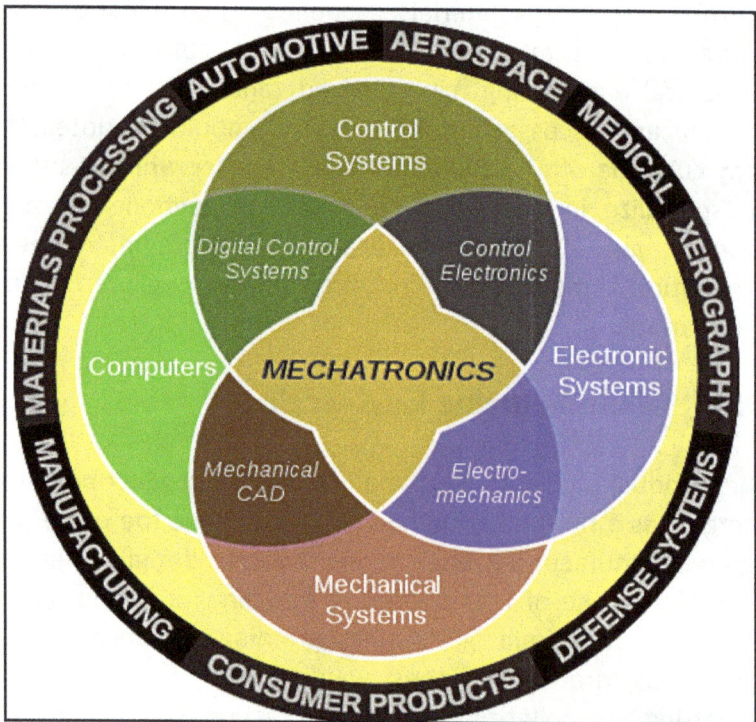

Figure 4.3. An Euler diagram describing the major branches of Engineering that make up Mechatronics *(en.wikipedia.org/).*

Mechatronics derives its name from its two core components: mechanics and electronics. Robotics/mechatronics engineering is a field of engineering which combines mechanics and electronics in designing and building machines that replicate human actions. Robots (humanoids) are electro-controlled mechanical structures designed jointly by mechanical engineers, electronic/control systems engineers, and software engineers. It is an emerging branch of engineering discipline and will become increasingly important in view of the exponential rise in deployment of artificial intelligence across the entire spectrum of human and social development. Robots are designed for industries such as mining, manufacturing and assembly lines, automobiles, military and law enforcement services and many more. Robots now carry out much of the traditional manual labor in manufacturing. For example, there are now at least as many robots as there are people in some modern automobile assembly plants. Robots are being used to remotely detonate explosives, explore ocean depths and terrestrial planets, drive vehicles, in crime investigations, in space exploration, and robot vacuum cleaners are becoming familiar features of the modern home while many others are assisting people with mobility issues. However, these robots still have to be designed, built, controlled by programming, and supervised, hence the increasing employment potentials of engineers who specialize in this area, particularly in the developed countries.

4.3.3   Production Engineering

Production engineering deals primarily with the planning of production processes: design of flowcharts, store and stock control, interfacing of processes, supply of materials to appropriate points at appropriate times, movement of semi-finished and finished products, planned maintenance, etc. The modern production plant can be very complex and malfunction of one unit or failure of supply of material to a station at the appropriate time can create a pile-up and possible shut down of a whole plant, with major financial consequences, hence the production engineer needs competence in quantitative planning tools such as scheduling, critical path analysis, queuing, stock control, quality control, etc. These tools are essential in order to operate a production system that minimizes downtimes and wastages in terms of precious time and materials, and optimizes the whole production setup. Most production engineers first train as mechanical engineers and complement with on-the-job training, but many institutions now offer undergraduate and graduate programs in production engineering.

4.3.4   Industrial Engineering

The industrial engineer is a systems integrator, a designer and manager of facilities which include building infrastructure (ventilation and air conditioning equipment, elevators, appliances, plumbing, etc.), individual and group workplace design, manufacturing and production systems, materials handling systems, service processes, production systems planning and control, resource allocation and scheduling, personnel assignment and scheduling, quality assurance, inventory control and system and personnel safety. Industrial engineers often have responsibility to design, oversee installations and manage large and complex systems or solve associated complex problems of existing systems, and therefore need strong grounding in basic engineering, operations management, quality control, statistical methods, and computer simulation. They will also work with other specialties and management, and need interpersonal communication skills. A typical industrial engineering curriculum covers basic engineering courses after which the student can specialize in one of many options including manufacturing engineering, production, engineering project management, management systems engineering, facility management, financial and costing techniques, business environment management, etc.

Industrial engineering offers the widest array of opportunities in terms of employment because of its flexibility and a very wide range of skill sets. While other engineering disciplines tend to apply skills to very specific areas, industrial engineers may be found working everywhere, from traditional manufacturing companies to airlines, from distribution companies to financial institutions, from major medical establishments to consulting companies, from high-tech corporations to companies in the food industry. The broad capability of the industrial engineer, strong skills in management principles, and close links with management provide enhanced opportunities for quick rise to management positions in complex enterprises.

4.3.5 Automobile Engineering

Automobile engineering is in fact mechanical engineering applied to the automobile industry: design and production of automobile engines, driving train, steering systems, braking systems, body structures, etc. The modern automobile plant is largely an assembly plant, with virtually all components outsourced to other mechanical production plants. Design of the engine and driving train, and testing and evaluation of the final product are the most challenging, and this is where the automobile engineer plays a key role. Assembly plants are now highly automated, many processes are carried out by robots, supervised mostly by technicians and production engineers but automobile engineers may also be involved, particularly in critical areas and when important changes are being made such as commissioning new assembly lines or introducing new technologies or models. Also, while most auto maintenance shops have mainly mechanics and fitters, many big ones have automobile engineers in charge. Modern automobiles now feature complex electronic, control and artificial intelligence systems, hence, although mechanical engineers form the core of automobile engineering, other engineering specialties are also closely involved, in particular, electronics.

4.3.6 Aeronautical/Aerospace Engineering

Aeronautical engineering is similar to automobile engineering, except that the knowledge of mechanical engineering is applied to aircraft: design of propeller and jet engines, fuselage infrastructure and fittings, landing gears, air-conditioning, etc. Again, the typical aero-plant is an assembly plant, with components coming from many sources. For example, American Boeing and the European Airbus plants receive around 80% of parts and supplies from over 12,000 sources from all over the world. Airplanes fly mostly in the first 6-12 kilometers of the Earth's environment which extends from the Earth's surface to about 100 kilometers. Space is beyond this height and home to many satellites, space shuttles and space stations designed for different purposes, including trips to the Moon and Mars. These equipment operate in different stress environments and feature much more complex equipment than aircraft, and therefore require more specialized engineering expertise currently evolving as aerospace engineering. The aeronautical/aerospace engineer is involved in the design, production and assembly of components, as well as testing and maintenance of aircraft, jet engines, rocket launchers, rockets, spacecraft, etc. A strong competence in mechanical engineering and aerodynamics is essential but, as in the case of automobile engineering, this specialty goes beyond mechanical engineering and has strong components of aero/astrophysics, electronic, control, and materials engineering.

4.3.7 Marine Engineering

The ship is a mechanical structure designed to operate in fluid environment, hence the marine engineer is a mechanical engineer who specializes in the design of floating structures. Also, most ships are driven by diesel engines, gas or steam turbines which are major products of mechanical engineering, while some, in particular military ships are nuclear-powered. Marine engineers specialize in the design of ships and boats and in the operation and maintenance of diesel engines in marine environment, operation of boilers, turbines, power generation equipment, maintenance of associated equipment such as propellers, turning gears, fuel storage tanks, ship's plumbing and air conditioning, etc. Some military ships and submarines are powered by nuclear reactors which generate heat energy for raising steam in boilers. Pressurized steam drives turbines which produce electric power required for driving propellers upon which ships depend for movement and manouvering, and for providing on-board utility electricity. While nuclear engineers manage the operation of the nuclear reactor, the marine engineer is responsible for the operation and maintenance of the steam boiler, steam turbine, driving gear and associated equipment, ship infrastructure and utilities. Marine engineers also train in communication and marine safety.

4.3.8 Agricultural Engineering

Agriculture was one of the first human endeavors to benefit from the emergence of mechanical engineering and has become heavily mechanized: tractors, planters, harvesters, processors, silos are normal equipment on a modern farm. The agricultural engineer is a specialized mechanical engineer with expertise in the design, operation and maintenance of farm machinery and associated structures such as barns, storage sheds, irrigation systems and equipment, etc. The agricultural engineer is also knowledgeable in soil science, soil management, crop science (seeding, tillage, harvesting), the design, production and operation of farming machinery (tractors, harvesters, etc.), processing flowcharts and machinery for farm products such as drying, pelletizing, pulverizing, storage. Training curricula include livestock and dairy production: poultry and fish farming and processing, plant/animal waste management, erosion control, water supply and irrigation, rural electrification.

4.3.9 Mechanical/Power Engineering

Electricity is produced primarily by using pressurized steam, pressurized hot gases, wind or water power to drive turbines coupled to alternators that produce electricity. Smaller units feature alternators driven directly by diesel or petrol engines. Power engineering involves production of heat from fuel combustion (or other sources such as agro-waste, solar, geothermal, nuclear) to raise steam for driving steam turbines and many industrial and domestic applications, the core of which is the steam boiler. It involves harnessing of water and wind energy or generating hydrogen to drive fluid turbines for power generation. It also involves the production of pressurized gases for the operation of pneumatic equipment. Design, operation and maintenance of the heat generator/steam boiler and associated equipment such as fuel combustion and materials handling equipment, solar and wind power equipment are the main responsibilities of a mechanical/power engineer, and thermodynamics, heat transfer, fluid mechanics, mechanics of machines, combustion technology are powerful tools at his/her disposal. The design, manufacture, operation and maintenance of solar farms, wind farms, wind, water, steam and gas turbines are within the purview of the power engineer, so are the design, manufacture and maintenance of the mechanical components of alternators that produce electricity. In effect, the mechanical/power engineer is responsible for around 80% of the design and operation of a typical power generation plant while the electrical section (alternators, transmission, distribution of electricity) is the responsibility of the electrical/power engineer.

4.3.10 Mechanical/Petroleum Engineering

It is not often obvious that mechanical engineering is at the core of petroleum production, from the design, operation and maintenance of drilling wells, on-land and floating drilling rigs, to pumping oil and gas and transportation to refineries, refinery and petrochemicals machinery design, installation, operation and maintenance, and pipeline distribution of products to consumers. Most petroleum production operations also operate captive power generation facilities, the management of which is within the purview of the mechanical (power) engineer. Also, many petroleum companies operate mini shipping facilities managed by mechanical/marine engineers. However, the process design and operation aspects of petroleum refining and conversion are managed by chemical or process engineers. Most mechanical engineering degree programs impart the basic knowledge requirements, complemented by on-the-job training, short training programs, or through the acquisition of postgraduate degrees in petroleum engineering. However, some institutions now offer undergraduate degree programs in petroleum engineering covering the whole spectrum of production, refining and conversion (a fusion of mechanical engineering and chemical engineering.

4.3.11   Mining/Minerals Engineering

Minerals are natural chemical compounds that are excavated from the Earth's sub-surface or deep deposits and processed into useful materials required in many engineering applications. Ferrous and non-ferrous ores are processed to obtain metals (iron and steel, copper, aluminum, lead, tin, zinc, etc.) while non-metallic minerals (limestone, clay, granite, etc.) are either used directly or, as in the case of limestone, converted into cement. Some minerals are in shallow deposits and can be recovered by removing the overburden (opencast mining) , while others may be located as deep as one kilometer, and underground mining techniques are required for recovery. Mining engineering involves the use of heavy machinery for ground surface excavation and tunneling underground, complex structures for underground roof support, and product evacuation by elevators and conveyors. A mining engineer is trained in basic mechanical engineering and specializes in the design of surface quarry; underground mining layout;

selection, operation and maintenance of heavy excavation and pneumatic equipment; design of underground mining and evacuation systems, equipment, roof support structures, mine de-watering, explosive blasting, mine safety systems. Mining engineers may also combine minerals processing which involves upgrade of ores through physical, chemical and thermal separation processes. Most minerals contain relatively high gangue (rubbish) and upgrading is almost always required. For example the iron content of an iron ore could be as low as 30%, and the ore has to be upgraded to 70% or higher to be acceptable as an iron blast furnace feed. The minerals engineer is a specialized mining engineer who has the knowledge and expertise to design beneficiation flowcharts featuring a series of appropriate mineral separation processes; select, install and maintain equipment; and supervise the overall operation of the plant. However, with some complementary or practical training, a graduate of chemistry, physics, mechanical engineering, materials engineering or chemical engineering can also function well as a minerals engineer.

4.3.12   Career Opportunities in Mechanical Engineering

Machines have become prominent features of virtually every human endeavor: production, manufacturing, agriculture, transportation, commercial, health, household, and the mechanical engineer is always in very high demand. Leading areas of employment opportunities are in machine building industries, manufacturing industries (machinery, automobiles, household goods, appliances, chemical, food, pharmaceutical), mechanized agriculture and food processing, power generation industry (fuel combustion, heat, steam, compressed air, electricity), fossil energy industry (coal, oil and gas production, transportation, equipment maintenance), hydropower, solar and wind power generation, transportation (building and maintenance of aircraft, ships, trains, automobiles). Other key areas of operation of mechanical engineers include mining and mineral processing, robotics, space rocket engineering, etc. Mechanical engineers also feature prominently in building construction and maintenance (heating, air conditioning, plumbing). Many mechanical engineers set up private businesses in mechanical component manufacture, as machinery installation and maintenance contractors, pipe network and communication mast installation and maintenance contractors, mechanical workshop and building maintenance services, heating and air conditioning engineers. Mechanical engineers feature prominently in sales departments of machinery building industries. They are also core staff in most research and development establishments because there is always the need for internal capability for the development, construction and testing of prototype machinery, installation and maintenance of mechanical equipment, etc.

4.4   ELECTRICAL ENGINEERING

Electrical power was invented in the later part of the nineteenth century, and quickly emerged as the power of choice for a very wide range of applications, from the operation of domestic equipment to the most sophisticated industrial manufacturing processes. In fact, it is difficult to imagine modern life without electricity. Direct current electricity was invented in the later part of the nineteenth century but it had severe limitations, in particular the restriction of distribution networks to only a few kilometers. The subsequent invention of alternating current electricity solved this problem and made it possible to step up electric power to high voltages in transformers at the generating point, transmit over long distances with minimal losses, and step down for local distribution to consumers. Electrical engineers design and oversee the manufacture, deployment and maintenance of alternators, transformers, high-voltage transmission grids, low-voltage consumer distribution networks, switching stations, electrical machinery and control systems. There are three major sub-specialties in electrical engineering as shown in Figure 4.4.

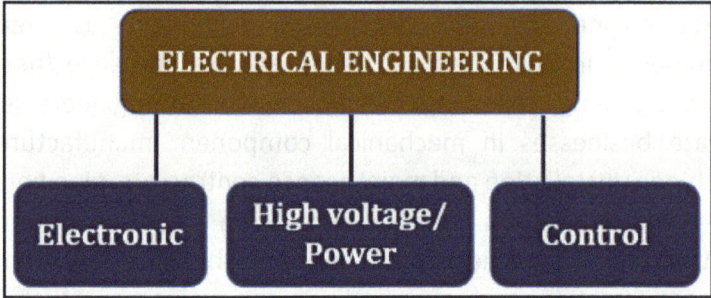

Figure 4.4. Sub-specialties in Electrical Engineering.

4.4.1   High-voltage (Power) Engineering

The work of the mechanical (power) engineer ends with the provision of a drive (steam, gas, wind, water turbine, or diesel engine) for the alternator which generates electricity, although the mechanical components of the alternator are usually designed and maintained by the mechanical power engineer. The work of the electrical engineer begins with design and winding of alternators, transformers for step-up of the current/voltage produced for long-distance transmission, and step-down for local distribution. The electrical (power) engineer has the competence to design, operate and maintain high-voltage power grids, low-voltage electrical distribution networks, major control stations and sub-stations. Competence also includes the selection and maintenance of the appropriate sub-station equipment and control systems. Furthermore, the very wide range of industrial and consumer electrical motors, tools, ovens, furnaces, power units for machinery, industrial and household electrical appliances, lighting are products of electrical engineering. The design and maintenance of the electric motor which is perhaps the most important electrical appliance is the joint responsibility of both mechanical (power) and electrical engineers. While the power (mechanical) engineer is responsible for the mechanical components (casing, rotor shaft, bearings, etc.), the electrical engineer looks after the electrical winding and control units. However, both have sufficient basic knowledge of engineering to quickly acquire the competence to maintain the two sections of the electric motor.

4.4.2 Electronic (Light-Current) Engineering

Electronics (light current engineering) is a more recent sub-specialty of electrical engineering which emerged with the invention of the transistor in the nineteen fifties, and the primary difference is in the level of voltage and current involved. While electrical engineers deal with systems that operate on thousands of volts, electronic systems hardly operate on more than 24 volts and currents no more than a few amperes compared with several thousand amperes in some electrical systems. The invention of the transistor revolutionized and miniaturized the design of a wide range of systems, from radio transmission to telecommunications and systems control, and prompted the emergence of the electronic engineer who has the expertise to design, produce and maintain complex electronic systems including microprocessors, smart phones, computers, medical diagnostic instruments, radio and television, the control systems of industrial machinery and domestic appliances, smart meters, and highly complex control systems featured in automobiles, aircrafts, spacecrafts, traffic control infrastructure, mechanical robots, driverless automobiles, etc. Micro-electronic modules and implants designed by electronic and biomedical engineers are also becoming prominent in medical care. The emergence of the television, computer, and more lately the smart phone as dominant consumer goods has further expanded the horizon of the electronic engineer, and the sub-specialty will likely lead the future growth of engineering as a profession.

4.4.3 Control Engineering

Developments in electronic engineering have opened up a new sub-discipline in electrical engineering: control engineering. Virtually every equipment that runs on electricity requires a control unit, from simple on-off switch and domestic remote controls to sophisticated remote controls for aircraft and space machines. Power generation, transmission and distribution, industry, transportation, communication, home appliances all depend critically on reliable electronic control systems. Electric power grid controls have become so sophisticated that power can be moved instantly to consumers within a nation and across borders in response to fluctuations in demand. Consumers' smart meters can be read remotely, while appliances in many smart homes can be controlled on line. Air transportation and space exploration would be near impossible without sophisticated control systems engineering. Also, it is inevitable that robots will become part of our daily lives, even in homes, and control engineering is at the heart of the emerging technology. Automobiles can now identify potential accident situations and take avoidance actions that are independent of the driver; airplanes can cruise and land on autopilot; many modern automobiles now feature auto cruise. The driverless auto-transport (cars, trams, trains) is another emerging development which depends critically on control systems. Remote-controlled drones now feature prominently in military warfare.

The typical early control system was only a bank of on-off switches, it was imprecise, labor-intensive, and subject to potentially tragic errors. Control engineering emerged with the electronic industry and deals with the design and deployment of control systems, from the simple television remote control to the sophisticated airplane, satellite and weaponry control systems. The monumental achievement of landing man on the moon in 1969 would not have been possible without control engineering. Sophisticated control systems keep the increasingly congested road, rail and air traffic safe and unmanned drones which are now potent military weapons could soon become a viable commercial transportation and goods delivery system. Artificial intelligence (AI) smart devices are at the core of control engineering. They can be programmed to monitor variables and adjust to conform with set values; they can collect and transmit   vital electronic information from equipment installations to control rooms, for example, from planet Mars over 50 million kilometers above to terrestrial control centers. Typical modern control systems monitor set voltages, temperatures and correct deviations, trigger remedial action or alarm systems. Modern road traffic controls, surveillance systems all depend on artificial intelligence.

Electric power grids are now linked across states and countries, thus facilitating international trade in electricity. It is not practicable or cost effective to shut down and restart power plants in response to fluctuations in demand and surplus power cannot be stored on any large scale. However, sophisticated control systems now allow power to be moved across grids instantly, which means that surplus electricity can be moved instantly from a location to other areas with high demand, often across state and country borders especially when there are significant time zone differences. The array of control systems (safety, entertainment, communication) is now one of the main selling points of the modern automobile. A modern home is full of small control units and programmed control systems can switch appliances on and off as desired, regulate temperatures, cooking time, etc., while smart systems can do same on line from thousands of kilometers away. The military depend critically on control systems for communications and guiding weapons that can hit targets thousands of kilometers away with incredible precision. Less human, machine-driven and robotic battlefield warfare systems are already under development, all of which will depend critically on control engineering.

A control engineer is basically an electronic engineer with focus on design of control systems, supervision of manufacture, and maintenance in service. A good grounding in software engineering is also vital since most smart systems need programing to function. Developments in robotics have posed new challenges for the control engineer: design and programming of controls for robots on assembly lines, automobile anti-collision devices, driverless vehicles, complex traffic light systems, surveillance systems, remote and robotic military warfare, etc. Micro electronic systems which are now being implanted in the human body, such as pacemakers and drug dispensing modules, robotic surgery, and medical diagnostic equipment all depend critically on control systems.

4.4.4.   Career Opportunities in Electrical Engineering

Electric power is indispensable in industry/manufacturing, maintenance, commerce, transportation, communication, health, education, residential sectors of the global economy, and opportunities abound for the electrical engineer: the design, operation and maintenance of power generation installations, complex grid distribution systems, industrial, commercial and household electrical machinery and appliances, etc. Electronic engineers are even more versatile because of the increasing dependence of every sector of the economy on electronic equipment: control systems, computers, the Internet, telecommunication, smart phones, household electrical appliances, etc. Healthcare is increasingly dependent on electronic equipment for diagnosis and treatment, including a variety of scans and electronic inserts such as heart pacemakers and drug dispensers. Opportunities abound in design, manufacture and maintenance of computers, complex control systems for spacecrafts, aircrafts, automobiles; appliances; telecommunication systems; design, operation and maintenance of computer and telecommunication networks, radio and broadcasting systems; medical diagnostics and treatment; and many other areas. Robots (products of mechanical and electronic engineering) are gradually displacing human effort in manufacturing and assembly lines, and advancements in the evolving electric and driverless automobile, drone and robotic warfare technologies will depend heavily on the electronic/control engineer. There are also very wide opportunities for electronic engineers in research and development, Internet commerce and technologies, the military and space exploration. Many electrical engineers also set up businesses in manufacturing, consulting and electrical services.

4.5 MATERIALS ENGINEERING

Materials have been prime movers of technology development from Early times and new inventions and innovations are possible only if the appropriate materials are available. The remarkable achievements in virtually all fields of human endeavor since the Industrial Revolution were made possible because the required materials became available: manufacturing, building, transportation, power generation, telecommunication, space exploration, information technology, medical diagnosis and treatment. In fact, non-availability of suitable materials is considered a major constraint to much more rapid technological development and innovations.

Most engineering disciplines focus on addressing human problems by constructing tools and shaping solutions. Materials science and engineering does so by studying, understanding, designing, and producing the materials those tools and solutions are made of, and on creating new materials that serve human needs. Materials engineering is a deeply inter-disciplinary field that encompasses nearly every form of matter: from producing large castings and forging complex steel structures to the atom-by-atom construction of nanomaterials and the controlled growth of biological substances. All engineering is grounded in science but, because of the relatively strong emphasis on science subjects (majorly physics, chemistry, biology, mathematics) most programs adopt the name 'materials science and engineering'. Furthermore, most programs allow for specialization in either the science (for example, bio and nanomaterials, See sections 4.12 & 13) or the engineering aspect which involves macro-materials production. Materials engineering deals with the family of materials that are indispensable to technology development: metals (iron and steel, copper, aluminum, tin, chromium, nickel, vanadium, columbium, silicon, uranium, etc.), non-metals (wood, ceramics and polymers), composites. Wood and ceramics (stone, and clay) have been in use from the Early times, so have a few metals that often occur in pure state in nature: gold, meteoric iron, copper. Composites (clay-fiber mixes) were also prominent building materials and rubber (a monomer) tapped from trees was in common use. However, the discovery of polymers (plastics) in the early part of the last century opened up a new field: polymer engineering.

Competence in the selection of appropriate materials for different applications and evaluation of environmental impact on materials properties have become very important areas of materials engineering. For example, prior to the Titanic disaster in the early part of the last century (in which a brand new cruise ship designed to be unsinkable cracked up on its first voyage), materials engineers were unaware that the strength of steel deteriorates rapidly as temperature of its environment drops towards freezing. Also, materials engineers have found that steel reinforcement of concrete structures such as bridge support pillars or pipelines embedded in saline soils and waters could corrode, weaken and disintegrate, often leading to disastrous accidents. These and many other events have opened up new areas of materials engineering such as alloy engineering and materials integrity in service. Materials engineers design and produce the materials that other engineers require to function and create new materials to serve new needs. The eight major sub-specialties of materials engineering are shown in Figure 4.5. Areas of competence of the materials engineer include design and operation of ore beneficiation, ore extraction (thermal, hydro, chemical) plants, primary materials production processes, materials forming and shaping plants and processes (casting, rolling, forging, drawing, etc.), finishing processes (chrome, nickel and tin plating plants).

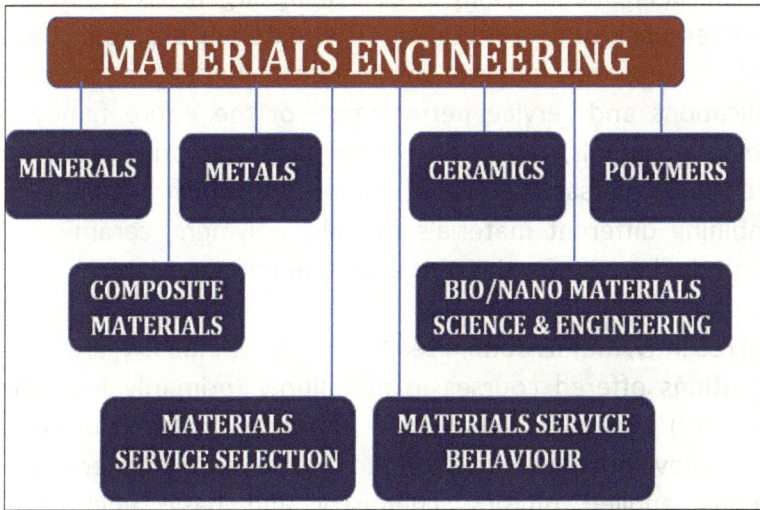

Figure 4.5.   Sub-specialties in Materials Engineering.

Relatively few metals are used in pure state because small additions of some elements can result in a radical change in mechanical, chemical, electrical, and thermal   properties and there are now thousands of steel, copper and aluminum alloys to choose from for specific applications. Alloy design and production have become prominent processes within the competence of the materials engineer. Selection of appropriate materials for different applications is also a major duty of the materials engineer, and requires a good working knowledge of mechanics of materials and materials-environment interaction.

New materials are being developed all the time to meet unusual requirements, for better service performance, or for new applications. In fact, the availability of appropriate materials is considered a primary regulator of (and often a major constraint to) technology advancement. For example, power could be generated much more efficiently at higher temperatures and pressures than are possible presently if suitable materials were available, and the pace of development in the electronic and biomedical engineering industries is being largely determined by the emergence of new materials. The materials engineer has a good working knowledge of production technologies, potential applications and service performance of the entire family of materials: metals, polymers, ceramics/refractories and the increasingly versatile group of composites which are made by combining different materials (metals, polymers, ceramics) to exploit the best properties of the component materials.

Until recently, metals dominated the materials industry, and most institutions offered courses in metallurgy (primarily industrial chemistry) and allied areas such as corrosion. Most of these courses have now been upgraded to metallurgical engineering by merging applied physics, chemistry and basic engineering courses. Developments in the refractory industry (used primarily in lining metallurgical furnaces and auxiliary equipment), extensive use of ceramics and glasses in industrial and domestic ware, the emergence of the polymer industry and exponential growth in use of composite materials have necessitated a change in scope of the metallurgical engineer, and many institutions now offer courses in materials engineering which provide competence in design, ferrous, non-ferrous, non-metal and composite materials technologies. It should be noted however that the division between various materials in not absolute: there are boundary overlaps, with some materials exhibiting properties between metals and non-metals, notably metalloids. Also, the group of non-metals is made up of solid, liquids and gases. The science of materials is a very strong component of training in materials engineering, hence the common composite name (materials science and engineering) of many departments.

4.5.1   Metals Engineering

Metals almost always occur in nature as complex oxide compounds and winning by reduction with carbon (charcoal) had been practiced for thousands of years (the humble beginning of metallurgy). Copper, brass, bronze, iron were produced and shaped into a wide range of products, from hunting and farming implements to weapons and ornaments. However these metals were being produced in very small quantities and increasing scarcity of hardwood for charcoal production was a major constraint. Scale-up did not occur until the Industrial Revolution when coal became available in commercial quantities and furnace temperatures were high enough to produce steel. Commercial production of non-ferrous metals also proliferated, in particular, copper, stimulated by the growth of the electrical industry, and aluminum in response to the needs of the chemical industry. About a quarter of the Earth's crust and three-quarters of all known chemical elements are metals and the material is dominant in the manufacture of industrial equipment and consumer goods. Also, although non-metals, notably sand, stone, wood, rubber have been in use from Early times and ceramic insulators became prominent with the emergence of electricity, metals were the primary engineering materials until the advent of polymers around the middle of the last century.

The main focus of the materials engineer is still on metals, in particular, ferrous (iron and steel), non-ferrous (copper, aluminum, tin, zinc), and the thousands of alloys that have been developed, from ore to product end-of-life, which may include recycling many times. Also, metals are now being combined with non-metals, notably polymers and ceramics as composites which seek to exploit the combined assets or compensate for deficiencies of the component materials. Metals are the primary materials in machinery, mechanical structures, electronics, pipelines, transportation, structural reinforcement, power generation and distribution, and in the manufacture of a very wide range of industrial and consumer goods. This explains why the study of metals still dominates most materials engineering programs. The main areas of metals engineering are extraction (thermal, chemical), shaping (drawing, rolling, forging, etc.), and finishing (surface hardening, plating, anodizing, etc.).

4.5.2   Ceramics Engineering

Clay has been in use from the Early times, but clay is a generic name that represents a wide range of oxide compounds, each with different composition and physico-chemical properties. The Early people knew which clay was good for pottery but not for building, they knew how to strengthen and glaze clay products for the production of fine works of art. However, developments in the metallurgical industry have helped to establish a new sub-discipline in materials engineering: ceramics engineering. Metallurgists use a very wide range of refractories (acidic or basic, high-temperature ceramics) for lining furnaces and ladles, and specifications vary depending on expected temperature regimes and the chemical properties of the molten metal. Ceramics are also being used extensively in other areas of engineering, in particular, as insulators and shields in electrical/electronic and nuclear engineering, building fittings and flooring, insulation, electronics, orthopedics, etc. There are not many natural ceramic deposits that can meet the currently stringent specifications required for most applications and the ceramics engineer is trained to produce a wide range of types to meet the requirements of different applications, by upgrading and mixing different ores and additives in appropriate ratios, appropriate processing, forming/shaping, and heat treatment to obtain the desired strength properties. Burnt ceramic bricks are also used extensively in building construction.

Glasses are often classified with ceramics because they are made primarily from sand (silicon dioxide) which is a basic ceramic raw material. However, while many ceramic products can exhibit considerable degree of crystallinity, basic glass is largely amorphous (non-crystalline) and exhibits properties associated with both solids and liquids. The history of glass dates back to around 3000 BC, based on the dating of glass artifacts found in Egypt and Syria. Glassblowing, forming, working, and coloring techniques emerged and flourished in the Mediterranean region over many centuries, largely for the production of artifacts and ornaments. However, the production of sheet glass (Sussex glass) in the thirteenth century AD marked the beginning of the emergence of glass as an important industrial material. Mass production technologies were developed during the Industrial Revolution, and glass science and technology as a research discipline emerged in the 1950s, pioneered largely by the automobile industry.

Hundreds of different types of glasses are now available, produced mainly by adding certain elements and compounds to molten silica to alter the crystal structure of the product and change the engineering properties. For example, aluminum oxide is added to improve durability, cerium is added to improve absorbance of infrared radiation, iron oxide and various other transition metal compounds are added to obtain desired colors. Soda-lime glass is by far the commonest type of glass in use today - for window panes, bottles and jars, domestic glassware, electric bulbs, etc. It is produced by melting a mixture of silica sand (SiO_2) soda (Na_2O), and lime (CaO), with various small additives to impart specific properties. The soda acts as a flux to reduce the melting temperature of silica while the lime acts as a stabilizer. In spite of the very broad utility of soda-lime glass, it has severe limitations: it is brittle and shatters easily, and it cannot withstand high temperatures. Addition of boric oxide (B_2O_3) greatly improves temperature resistance and makes it possible to use glassware in ovens and microwaves (for example, Pyrex products). Ceramic glasses capable of withstanding temperatures as high as 1000°C are now available for a wide range of industrial applications. In summary, glass science and engineering has now become a major area of specialization in ceramics engineering.

4.5.3 Polymer Engineering

The development of plastics started several hundred years ago with the use of natural materials that had intrinsic plastic properties, such as chewing gum and shellac. Subsequently, natural materials such as rubber, nitrocellulose, and others were being chemically modified and used extensively. Natural plastic is made up of simple chains (monomers) of covalently-bonded carbon atoms and other elements, in particular, hydrogen and oxygen. They are pliable and easily shaped, hence the name 'plastic'. In the last part of the nineteenth century, several scientists synthesized various types of plastics with longer and more complex arrays of monomer chains, hence the name 'polymer'. However, the first major breakthrough was the invention and commercialization of fully synthetic bakelite in 1907, as a more effective replacement for natural shellac (a versatile natural polymer made by the female lac beetle) which was then in common use as electrical insulator. This development and ready availability of petroleum (an abundant carbon source) stimulated major investments in polymer research, and a wide range of polymers with different properties soon became available, including nylon, terylene, plexiglass, polyethylene, polystyrene, polyurethane, polyvinyl chloride (PVC), to name a few. The nature of covalently bonded carbon chains, length of chains and the patterns in which they are arrayed make polymers strong, lightweight, insulating, flexible and highly unreactive, and the possibilities for use are almost endless. Over time, plastics have replaced traditional materials in many applications including packaging, water and gas distribution, automobile components, insulation, household goods, clothing, furniture, food packaging, computers, cell phones, biomedical inserts, toys, etc. The polymer engineer is trained to manipulate carbon chains obtained mainly from petroleum in order to obtain a wide range of desired properties, and also to form into useful products.

4.5.4   Bio/Nano Materials Engineering

Biomaterials (materials that are designed to interact with biological systems for a medical purpose) have been in use for over a century, particularly in dentistry and treatment of bone fractures. However, the scope of applications has increased exponentially in the last fifty years or so, and a wide range of devices are used as therapeutics to treat, augment, repair or replace a body part or tissue, or in medical diagnosis. Biomaterials are either natural materials or synthesized in the laboratory from polymers, ceramics or composite materials, many on micro or nano scales. Examples include heart valves, heart pacemakers, tissue implants, hip implants, joint replacements, bone repair, cosmetic implants, contact lenses, dental implants, drug delivery capsules, and many more. Current research is extensive and includes development of fully functional artificial human heart and kidney. The human body is hostile to all foreign objects such as implants, and one critical requirement is that the material must be biocompatible to minimize complications or rejection. This explains why polymers and ceramics which are highly unreactive are becoming materials of preference in biomaterials engineering. This relatively new field is growing at a phenomenal rate and requires integrated training involving elements of materials science, materials engineering, biology, medicine, chemistry, and tissue engineering.

Chitin is a natural polysaccharide biopolymer and the second most widely distributed after cellulose. It is the building material that gives strength to the exoskeletons of crabs, shrimps, insects, the cell walls of fungi, and is a byproduct of the food processing industry. It can be converted into chitosan, a potentially versatile biomaterial by biosynthesis. Chitin and chitosan are both biocompatible, biodegradable, and non-toxic natural biopolymers. They are also antimicrobial and hydrating agents. Owing to their unique biochemical properties such as biocompatibility, biodegradability, non-toxicity, ability to form films, etc, these materials have found many promising biomedical applications (Elieh-Ali-Komi, 2016). These developments are discussed in some depth in Section 4.13.

Nanomaterials engineering involves atom-by-atom fabrication of materials and is currently one of the most researched areas of materials science and engineering. Recent innovations include the development of carbon fibers, ten thousand times thinner than a human hair, yet a hundred times stronger than steel. The areas of potential application are almost limitless. Nanocarbon fibers embedded in polymer, ceramic or metal matrix account for around two-thirds of fabrication materials for modern airplanes and spacecrafts. Other products include lightweight bulletproof vests, sporting equipment, orthopedic implants, and many other products that require high strength-to-weight ratio. Nanofiber-reinforced ceramic materials are used as heat shields on spacecraft, and nanoscale biomaterials are used in medical implants and drug delivery. Developments in nano science and technology are opening up new areas of competence in bio and nano materials engineering: design and fabrication of biomaterials and implants on the micro/nano scale for different applications.

4.5.5   Minerals Engineering

Most metal ores have low metal content which could range from less than 1% for tin and uranium ores to around 35-60% for iron, and upgrading is almost always necessary before smelting. For example, most steel plants now require at least 70% Fe ores to feed blast furnaces and uranium U-235 fuel for nuclear plants must be upgraded to at least 6%. Also, copper ores which may be as lean as 5% Cu have to be upgraded in a series of complex processes to obtain the 99.99% Cu purity required for many electrical and electronic applications. Mineral engineers specialize in the design and operation of mineral processing technologies and core areas include beneficiation processes (physical separation processes such as comminution, flotation, concentration and dewatering), and extraction methods (chemical methods such as bio/hydro-extraction; chemical leaching;   electro-metallurgy;   and   thermal   extraction technologies). Very lean ores such as copper, tin and uranium ores require specialized processes and equipment for upgrading and, apart from complexity, processing of some ores such as uranium can be extremely dangerous and mineral engineers need special training for such operations. Course programs also include some basic engineering courses, physics and chemistry of mineral ores, unit operations, thermofluids, transport processes process/plant design and operation, and process optimization. While some institutions offer degree programs in minerals engineering, graduates with qualifications in many other areas of science and engineering can convert through retraining or in-plant experience.

4.5.6   Material Systems Engineering (Materials selection for service)

This is a very important area of materials engineering because many often conflicting factors need to be reconciled in choosing a particular material for a given application: strength specifications, service environment, manufacturing processes, etc., and strong competence in both materials science and engineering is required of a materials selection engineer. For any given metal there is a very wide variety of potentially suitable metal alloys; also, there are hundreds of polymers, glasses, ceramics and composites to choose from. The materials systems engineer acts as a prime liaison between the materials industry and potential customers who often know a lot about their design but very little about the appropriate materials. Material systems engineers are major sources of information about the materials' needs of industry, often leading to the development of new materials. They also play an important role in forensic investigations of materials' failure in service. A materials engineer can specialize in this area by acquiring additional knowledge on the technical properties and environmental compatibility (stress, temperature, corrosion) of the large number of metal alloys, ceramics, polymers and composites that are currently available.

4.5.7   Materials Integrity Engineering (Materials behavior in service)

Metals, in particular, steel are unstable because they have been forcibly deprived of oxygen, and tend to revert to natural stable oxides when exposed to oxygen (oxidation), or react with other chemical compounds in aggressive chemical environment to form more stable compounds (corrosion). These reactions can degrade the engineering properties of steel significantly, and even cause failure in service. Degradation may be greatly enhanced by the nature of environment to which the metal is exposed and the prevailing stresses and strains. The service integrity of an oxidized or a corroded metal is greatly compromised, hence metallurgical engineers go to great lengths to prevent or minimize metal degradation in service, including plating and alloying, sacrificial alloy protection. Protection may slow down metal oxidation and corrosion but they are still major problems in engineering. Steel in reinforced concrete immersed in saline water will eventually corrode and lose strength and buried steel pipes will eventually corrode and fail especially when also subjected to stresses. Soldered electronic components often corrode and even plated metals may eventually fail due to micro cracks through which oxygen and corrosive fluids can penetrate.

High alloy steels and aluminum are perhaps the most resilient utility metals but both may also corrode and fail if the conditions are sufficiently severe. The annual cost of corrosion worldwide is estimated to be over US$5 trillion, equivalent to 5 – 6% of the global GDP (one trillion in the United States alone). Chemical, petrochemical, food industries and water supply systems are particularly vulnerable, but substitution of metal with plastic in pipes and storage systems has greatly reduced the problem. Furthermore, metal cans for foods, drinks and chemical storage are now lined with polymers in addition to plating. The materials integrity engineer is trained in metals engineering, metals-environment interactions, principles of materials selection for applications, materials failure investigative techniques, degradation preventive measures, and non-destructive testing. Graduates of materials science and engineering have the basic training to function well as materials integrity engineers but graduates of other engineering specializations or physicochemical sciences can also retrain and function well in the field.

4.5.8   Career Opportunities in Materials Engineering

Advances in materials science and engineering largely determine the pace of technology development. Without the invention of the silicon transistor and subsequent development of integrated circuits and microprocessors, the major advances in the past few decades in electronics, computer engineering, telecommunications, transportation, and control engineering would not have been possible. Materials engineers have developed thousands of alloys, polymers and composite materials for different applications, including strategies for prediction of behavior in service. Materials failure in service have been the primary causes of many disasters, such as the sinking of ships, air and space disasters, failure of bridges, etc., and end-users depend critically on the advice of materials engineers in choosing appropriate materials for different applications, and monitoring performance in service.

Materials engineers work in primary metals industries (iron and steel, non-ferrous metals), metalworking and coating industries (rolling mills, forge presses, foundries, plating shops), mineral processing industries, ceramic industries, advanced materials industries (composite, bio and nanomaterials industries), electronic industries. They also work in manufacturing, chemical, petroleum, petrochemical, food, water, ship building and many other industries as corrosion monitoring and prevention engineers. Materials engineers also play a prominent role in forensic investigation of accidents such as failure of bridges, aircraft engines, space shuttles, etc. For example, material failures have been identified as the primary cause of many disasters; failure of aircraft engines in flight which occurs fairly frequently is often due to the fatigue fracture of a turbine blade or other critical engine component; and many equipment failures in service are caused by stress corrosion cracking of vital components. A lot is learnt in the investigation of every materials-related disaster which helps to stimulate the development of new materials. Furthermore, materials' constraints are holding back the commercial exploitation of many new technologies including high-efficiency power generation, electric power storage and nuclear fusion. The world is looking up to proliferation of electric vehicles as a major mitigant for environmental pollution but the procurement of vital rare earth metals for batteries which account for around 40% of the cost of the vehicle is becoming increasingly problematic. The field of materials research is one of the most dynamic areas of technology research and development globally, driven by exponentially growing demand for new materials for a wide range of applications, from power generation to biomedical applications. A postgraduate degree in any area of materials science and engineering opens up a much wider range of employment opportunities for materials engineers, especially in the very dynamic areas of research and development.

4.6   CHEMICAL ENGINEERING

Chemical engineering is the specialization area of engineering that deals with the design and manufacture of chemical and biological products: transportation fuels (petrol, diesel oil, aviation fuel), gas, chemicals, petrochemicals, fertilizers, polymers, food, pharmaceuticals, soap, detergents, lubricants, and many household products. Chemical engineering was introduced in some institutions in Europe and the United States towards the end of the nineteenth century as a merger between chemistry and engineering subjects, and suffered identity problems for decades. However, from the early 1900s, it emerged as an independent specialization in engineering prompted by the rapid developments in the chemical, petrochemical, pharmaceutical, and food industries. All these industries rely primarily on a number of key engineering unit operation processes, (notably filtration, separation, evaporation, polymerization and crystallization), and transport phenomena (interactive treatment of mass, heat and momentum transfer). These subjects form the core of a typical chemical engineering degree program. A strong flare for chemistry is required, in addition to mathematics and physics: while it is possible to gain entrance to mechanical or civil engineering programs with chemistry at ordinary level, a chemical engineering aspirant requires advanced level chemistry. Many chemical engineering courses have a common 2-year core engineering program, with specialization in the third year, but many also start specialized and unique courses from the second year.

A very large number of industries depend on the synthesis and processing of chemicals and materials and chemical engineers have the competence to work in refineries, natural gas processing plants, chemical, petrochemical, food, pharmaceutical, minerals processing cement, primary metals industries, and many more. One unique feature of chemical engineering is the process commonality in the different types of industries. It means that a chemical engineer with good grounding in chemical reaction engineering, unit operations and transport processes can quickly adapt and function in any of the key chemical industries. The main sub-specialties are shown in Figure 4.6.

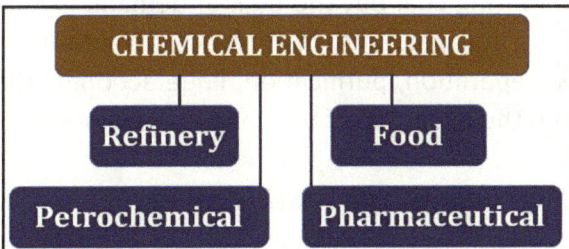

Figure 4.6. Sub-specialties in Chemical Engineering.

4.6.1   Refinery Engineering

Petroleum engineering became a distinct sub-specialty of chemical engineering with the discovery of oil and gas in the early part of the last century. Refineries convert crude oil into transportation fuels through a series of unit operations, producing petroleum naphtha, gasoline, diesel fuel, asphalt base, heating oil, kerosene, liquefied petroleum gas, jet fuel and lubricants. Some of the by-products of the refinery are feedstocks for many other major industries including polymers, chemicals, pharmaceuticals and fertilizers. Asphalt which is used extensively in road paving and corrosion protection is a byproduct of petroleum or tar sand distillation. Chemical engineers are familiar with the theory behind most refinery processes and with some practical training, they can operate as process engineers in refineries.

Natural gas is also processed to strip it of undesirable constituents that inhibit its direct use as an industrial or residential fuel, notably sulphur. Therefore, crude natural gas is passed through a separation process to recover some compounds which are more valuable as petrochemical feedstocks (natural gas liquid or NGL), stand alone fuels, or industrial gases notably propane, ethane, butanes, helium. The clean gas may be supplied to consumer pipelines, or liquefied for storage or transportation, while the separated compounds are processed into valuable products including propane gas, polymers and petrochemicals. In some countries, coal gas is produced for industrial and domestic use by carbonizing coal but the bulk of demand is now filled by processed natural gas. The main processes involved in these operations (separation, purification, liquefaction, carbonization) are all within the expertise of the chemical engineer.

4.6.2   Petrochemical Engineering

Apart from transportation fuels and other refinery products, many important organic chemicals are also   produced from petroleum. The main feedstocks are olefins and aromatics obtained from naphtha, gas oil from oil refineries, or natural gas oil from gas processing plants. Synthetic gas obtained from coal or by-product gas from coke ovens or coal gasification is also a suitable petrochemical feedstock and, in some cases yields better quality products. Depending on the feedstock which is cracked at high temperatures, some major chemical building blocks (ethylene, propylene, butylene and butadiene) are produced in different ratios. Primary petroleum feedstocks may also be polymerized to form synthetic resins. These chemicals and resins are precursors to a very wide range of products: plastics, soap, detergents solvents, lubricants, paint additives, fertilizers, explosives, pesticides, synthetic fibers and rubbers, drugs, etc. Chemical engineers are the prime movers of the petrochemical industry in view of their competence in the major processes: cracking/catalytic cracking, isomerization and reforming.

4.6.3   Food and Pharmaceutical Engineering

Production of food, beverages, drinks, drugs involves a series of unit operations: size reduction, drying, mixing, separation, filtration, steam processing, compaction, packaging etc., all within the competence of chemical engineers (also called process engineers). They also design processing, handling, and packaging equipment. Traditionally, mechanical, agricultural, and chemical engineers, or biochemical scientists were retrained to operate in food industries but many institutions around the world now offer courses in food engineering.

4.6.4   Career Opportunities in Chemical Engineering

Most chemical engineering programs adequately equip their graduates with the theoretical basis they need to function as process engineers in refineries, gas processing and liquefaction, and any chemical, petrochemical, food or drug manufacturing industry, and most gain practical working experience on the job or through short training programs. Considering the core status of fuels, petrochemicals, food and pharmaceutical products in the world economy and human development, career prospects in chemical engineering are extremely wide. Chemical engineers can also work in minerals processing, metals extraction and refining, electroplating, biofuels production, cement industry, or as corrosion engineers. Opportunities in consulting, entrepreneurship, education, research and development are also very wide.

4.7 COMPUTER ENGINEERING

The first computer as we know it today was an analytical engine invented by Charles Babbage in the nineteenth century AD. It was a crude design but it established a basic framework which has become a standard for computers of today. The first major advance came in the early 1940s when the first electronic computer (the Colossus) was built for the US military. This was followed rapidly by the emergence of a general-purpose computer, the ENIAC built in 1946. It employed around 20,000 vacuum tubes and weighed about 30 metric tons (Figure 4.7).

Figure 4.7. The ENIAC (1946), the world's first programmable, general-purpose electronic digital mainframe computer.
(https://penntoday.upenn.edu).

The computer could only perform a single task at a time, it had no operating system, and input and output were by punched cards. Today's average laptop computer is more powerful by several orders of magnitude than the ENIAC. However, it carried out calculations in 30 seconds which took human 20 hours.

The first major revolution in the computer industry came with the invention of the transistor which replaced vacuum tubes, miniaturized the computer, upgraded its capability, and made possible the design of the first commercial computer in 1951, followed rapidly by several improved models. However, they were all huge systems with relatively low computing power and use was restricted to the military and large industrial, commercial and research organizations. Another major revolution in the industry came in the early 1960s with the invention by United States Texas Instruments of the integrated circuit, also called a microchip or microprocessor in 1958 (a semiconductor wafer on which up to millions of tiny resistors, capacitors and transistors are fabricated and networked) which can function as amplifiers, oscillators, timers, computer memory, microprocessors, etc. This made possible a major size reduction, increased power, multi-tasking capability, massive data storage, development of the personal computer, and the emergence of computer engineering as a distinct profession.

Computers (hardware) can only operate on input instructions in binary language from the user, hence the need for an interpreter: software/operating system which required the input of a series of instructions from the user and was beyond the capability of ordinary people. Different operating systems and software packages were developed but were still too complex for general use. The emergence of the Windows operating system in 1983 resolved this problem and reduced instruction input to clicking on icons. Without this development, it is difficult to imagine the development of the personal computer and the cell/smart phone which now form the core of personal possessions all over the world. Over time, computer engineering has evolved as three distinct but mutually dependent sub-specialties: hardware, software and network engineering (Figure 4.8). Several other sub-specialties are also emerging, notably systems engineering and cyber security.

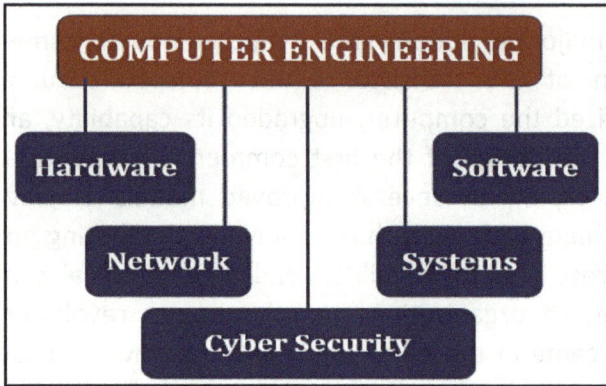

Figure 4.8. Sub-specialties in Computer Engineering.

4.7.1   Computer Hardware Engineering

Hardware engineering is the aspect of computer engineering that deals with the design of computers from very simple systems (for example, calculators, children's toys), to systems that are capable of performing a wide array of functions, such as those used to control national power grid systems, air and road traffic, airplanes, space satellites, space shuttles, etc. Computers also dominate operations in commerce, design, manufacturing, education, health, entertainment, communications, industry, robotics, military operations. Over 80% of the world population currently own smart phones which are products of computer engineering. Computer hardware engineers design, develop and test computer components such as microprocessors, (chips), integrated circuit boards, memory devices, input-output devices, data storage devices, networking components, and other associated devices; they integrate these numerous devices into computer and network systems; and they design mini/micro computer systems for general or specific applications, such as smart phones, watches and TVs; automobile brain boxes; electronic surveillance and traffic control systems; driverless control systems; implantable biomedical micro-computers, etc. The hardware engineer takes credit for miniaturizing computers: the earliest computers filled large rooms, yet had very low computing power compared with today's laptop or even cell phone. An incredible number of manufacturing processes, machinery and consumer products today are controlled by mini computers. Automobiles, telecommunication systems, traffic control systems, aerospace machines, industrial equipment, utilities, medical diagnostic equipment, biomedical implants, education systems, appliances, children's toys, and many more depend critically on computers or microprocessors. A strong flare for electronic engineering is the main prerequisite for a career in hardware engineering (see Section 4.4.2).

4.7.2   Computer Software Engineering

A computer understands and speaks only binary language comprising strings of zeros and ones, it is of little use unless it receives instructions in the appropriate language, and hundreds of languages have emerged in the last five decades or so. Software engineering is about the development of programs which act as translators between the computer user and the machine. In effect, computer instructions including commands, images and sounds however complex and in whatever non-binary language must be converted to a string of binary codes, also known as software. The earliest languages required input of a large series of codes punched on tapes, and use was beyond the competence of ordinary people. Further development produced the personal computer (PC) and also led to the emergence of the Disc Operating System (DOS) which still required that the user typed a series of memorized complex instructions. In effect, use of the PC was still very restricted. The real breakthrough came with the development of Windows, an operating system that requires the user to do little more than move a mouse around, clicking on icons. What is in fact happening is that software engineers have done most of the work by writing detailed programs and instructions linked to the icons, assigning a pattern of binary digits also known as bits to each letter or character that makes up instructions. This development transformed the computer into a universal tool that can be operated by almost anybody; it revolutionized the development of many other products, in particular, the smart phone and Internet commerce.

The emergence of the user-friendly touch screen technology which is a result of joint effort of the hardware and software engineers, has further simplified human interaction with computers, and voice control of computers, televisions and smart phones   (speech recognition) is already a reality. Software engineering (also known as programming) has produced a very wide range of software packages for key operations such as word processing, spreadsheets, book keeping and accounting, database management, computer-aided design, computer-aided manufacturing, smart control systems and device drivers, smartphones, education, healthcare, banking, weather forecast, media games, personal accounting, antivirus, spyware, firewalls, encryption, personal wellness, virtual games, dating, social media, and many other applications. In fact, an 'app' is now available for almost any enterprise (businesses, sports, religious groups) or indeed any conceivable human endeavor. The emerging artificial intelligence (AI) and robotic technologies also depend critically on programming.

4.7.3   Computer Network/Systems Engineering

Computers can operate as stand-alone or as part of a complex network. Internet browsers and search engines, Internet communication and commerce, telecommunication control centers, education, health, utilities, space exploration, air traffic control, weather forecasting, security, all depend critically on networks of up to many thousands of computers located all over the world. Network engineers are basically hardware engineers who specialize in designing system configurations, directing system installations, establishing operation procedures, managing and maintaining computer networks. They are experts in the selection and deployment of appropriate servers, routers, switches, cables, storage systems, etc. While they do not usually write programs, they have good working knowledge of networking software. They also have responsibility for maintaining the integrity and security of networks and connected systems. Network engineers are the brains behind the Internet, search engines, browsers, e-communication, e-commerce, multinational business operations, social media, etc. Also, many organizations whose activities may spread across countries and regions of the world: industry, commerce, education, healthcare, advertising, entertainment, social networking, research are now closely inter-connected through complex, Internet-based computer networks. They are migrating to cloud computing which facilitates sharing of software and access to data from anywhere in the world.

4.7.4   Cyber-Security Engineering

The Internet and World Wide Web facilitated by computer networks have revolutionized and globalized communication, reducing interaction between two farthest parts of the world to a few seconds. They have also facilitated business, trade and commerce across the world. This is possible because systems all over the world are now inter-connected through the Internet, including hardware, software and data, making them potentially exposed and vulnerable to unauthorized access, also known as cyber-attacks. Digital information and data processing, storage and transmission are already at the core of most modern enterprises and most individuals connected to the Internet have significant digital footprints. Computer networks and information systems operating in cyber-space (interconnected on the Internet) are at the core of modern businesses many of which operate across countries and continents. Government and human development enterprises (health, education, social services, utilities, security, etc.) and commerce depend critically on Internet-based operations. The traditional systems of in-house applications and data storage are rapidly being replaced by shared or independent Cloud services. However, these highly beneficial developments in information technology also come with a variety of cyber-threats. Some attacks (viruses) are simply malicious, with the primary intention of corrupting or destroying databases, while others target harvesting of valuable information and data such as personal data, account details, proprietary information for financial benefits. Cyber attacks which target enterprises, can compromise millions of personal data, or cripple operations across continents are now fairly common occurrences.

The risks of exposure to cyber-attack may originate from personal cyber-habits, employees, clients and contractors, or external cyber-criminals; they may result from deliberate acts or human errors. Irrespective of the source or cause, the consequences can be devastating, ranging from valuable or sensitive data or financial loss, to massive disruption of operations of critical infrastructure. Cyber-crime is increasingly weaponized across countries to extract ransom payment, cripple sensitive infrastructure, meddle in the affairs of opposing groups, or the politics of enemy nation states. Cyber-security has emerged as a major technology discipline and, with the exponential rate of personal and corporate migration to cyber-space, incidents of cyber-crime are projected to grow at a similar rate. Every computer, database, smart cell phone that is connected to the Internet is vulnerable and now needs protection. While anti-virus programs may be sufficient for personal computers, extensive networks and databases such as for banks, government, military, industrial and commercial enterprises, health facilities, institutions require elaborate protection and the exponential rate of cyber attacks especially in the last few years has brought the evolving discipline of cyber-engineering into prominence. Cyber-security engineers have emerged as indispensable members of computer systems teams in most enterprises but many also rely on independent security businesses. Competence is required primarily in software engineering but working knowledge of hardware network engineering is also vital. There are currently relatively few degree programs in cyber-security engineering but many computer engineering programs include courses on cyber-security and some offer specialization options. Also, scientists, computer engineers, programmers can retrain through short, specialized, product-specific courses offered by software vendors, or diploma/graduate courses offered by many colleges.

4.7.5 Career Opportunities in Computer Engineering

The computer is at the heart of everything, from space shuttle control, to children's toys. Almost any business can now be created and completed on line: commerce, healthcare, education, job search and interviews, games, love dating, even cyber-warfare. The fact that the Internet/World Wide Web is timeless and available round the clock is a major attraction for both the product provider and the customer. Marketing and operational costs are greatly reduced; potential customers are more diverse and better targeted, with the role reversed - the customer now looks for the provider rather than vice versa. Internet commerce is now global and exposes products to virtually every corner of the world, with improved After-Sales-Service. Retail e-commerce around the world is currently around $3 trillion and is increasing by around 25% each year. Online shopping for almost anything: household consumer goods, grocery, services, delivered to doorsteps is growing exponentially and gradually driving traditional shopping to extinction, driven largely by the increasingly busy household schedules and the convenience of shopping at home without time constraint.

Virtually every major enterprise - industry, banking, commerce, human development services, transportation, utilities, communication, research and development - has its own computer network, and Internet commerce facilitated largely by computer networks is taking over from traditional high street shops all over the world. Shop purchases are paid for by swiping a card or touching a smart box with a smart phone; in some new systems, checking out from a grocery is no longer necessary because every item placed in the cart is automatically debited to the customer's account; the traditional banking system is rapidly becoming obsolete and many branches have closed because it is no longer necessary to visit banks to make most transactions which can now be done on line. Most people now spend their money without going to the bank or handling currency, and money can be received or paid electronically, thanks to the network and software engineers.

Around two thousand satellites currently rove the skies transmitting communication all over the world, and an estimated 50-60,000 will be launched over the next few years, all controlled by earth stations which rely primarily on electronic and computer engineers. Apart from transmitting communications, they map the world; they guide human, aerial, and marine transportation (GPS); they provide critical data for weather forecasting; they are indispensable in national security and military operations. Programmers are in very high demand in all areas of computer deployment and computer network systems engineers are engaged to develop new networks, websites, processes and procedures, monitor, maintain and improve existing ones, or resolve problem areas. Many programmers also operate independently, developing potentially useful applications which are eventually sold to end users. Microsoft, Yahoo, Google, Amazon, Ebay, Whatsapp, Twitter, Facebook, Skype, Zoom and many others are all products of entrepreneurial computer engineers. Around 5 billion people all over the world now have cell phones loaded with a very wide variety of software, and are permanently linked to telecommunication stations and the Internet. The rapidly increasing global dependence on computers, cell phones and control systems means excellent prospective employment opportunities for graduates in any area of sub-specialization of computer engineering. Virtually every industry, financial institution, educational institution, healthcare facility, commercial enterprise, utility industry, maintains a computer network and carries out substantial business on the Internet, managed by computer engineers. Around half of the world now literally lives on line: shopping, banking, socializing, communicating, controlling homes on line, etc., and, inevitably, cyber-crime is also growing exponentially. Cyber-security is increasingly becoming a major global issue and computer hardware and software engineers are at the fore front of this growing area of specialization. The average home in the modern world now has its own mini network of interconnected routers, computers, cell phones, and other electronic devices and depends on independent computer businesses for supply and maintenance, thus creating wide opportunities for entrepreneurship.

4.8   SYSTEMS ENGINEERING

Systems engineering was developed in the 1950s as a strategy for analyzing complex defense systems but has become a very versatile and powerful tool for analyzing existing systems or for the development of new systems from concept to operation. It is applicable to all kinds of project developments or improvements, from building construction to design and implementation of complex transportation systems, power grid distribution systems, other utilities, and indeed any system in which many different activities have to be integrated and synchronized. It starts with full definition and documentation of requirements, identification and full specifications of all the elements that make up a system, interactions between system elements, potential interfacing problems and issues. A prototype design is then developed by synthesizing elements, evaluated, tested and refined until a reliable quality product that meets the desired goals is achieved. Systems engineering is interdisciplinary and different specialties are usually integrated into a team. The team leader is usually a systems engineer who is competent in at least one of the required specialties but, more importantly, has leadership and management skills to coordinate the work of a team made up of different specialties. Computer skills are indispensable since projects can be very complex and continuous scheduling and progress analyses by the team leader are vital.

4.9   TELECOMMUNICATIONS ENGINEERING

Telecommunications engineering evolved as a specialty from the late nineteenth and early twentieth centuries AD in response to developments in the telegraph, telephone, radio and television industries. The telecommunications engineer is responsible for the design, installation and maintenance of network infrastructure, allowing for inter-connectivity of devices and people: cable networks, wired and wireless transmission stations, digital satellite communication systems. Other duties include design and installation of radio and television broadcasting stations, mobile broadcasting units, static and mobile transmission and receiving stations. Telecommunications engineers are the prime movers of Internet and wireless communication technologies which have revolutionized all aspects of human development: commerce, banking, social media, data management, education, healthcare, etc.

The whole world is now being linked by ocean-immersed cables (backbones), underground cables and satellites. Current upgrade from electrical cable to much more efficient and versatile optical fibers is one of the latest major innovations in the telecommunications industry; the fifth generation wireless technology (5G) which is currently being developed and deployed means much faster downloads, outstanding network reliability, and greatly improved performance for Internet users. Developments of this new technology upgrade and all the earlier G1-G5 communication systems have been pioneered by telecommunications engineers. Satellite communication technology deployment is growing very rapidly and is bringing Internet connectivity to previously deprived communities all over the world. Telecommunications engineering is one of the world's fastest-growing engineering specialties, offered mostly as postgraduate programs for electrical, electronic and computer engineers, all of who already have core competence and can quickly acquire complementary theoretical and practical knowledge of modern communication systems: telecommunication systems design, IP addressing, modulation/demodulation, channel coding and decoding, multi-user communication systems, etc., as well as evolving new technologies such as fiber-optic communication and environment simulation. Some institutions also have undergraduate programs in telecommunications engineering. The specialty is a fusion of electronic and computer engineering and three sub-specialties have emerged as shown in Figure 4.9.

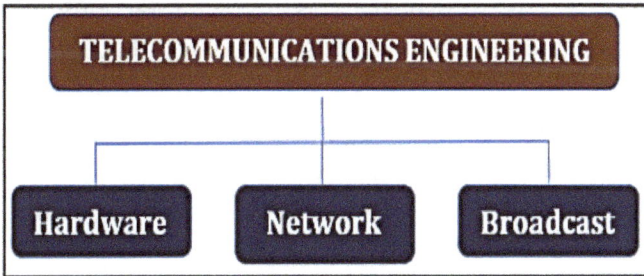

Figure 4.9. Sub-specialties in Telecommunications Engineering.

4.9.1 Telecommunications Hardware Engineering

This sub-specialty involves design, manufacture, operation and maintenance of telecommunications equipment: transmitters, receivers, aerial masts, cable, fiber, microwave transmission equipment, switching systems, multiplexers, equipment for communication satellites and earth stations, and other specialized equipment associated with telecommunications networks. The hardware engineer may also be involved in the design and installation of external infrastructure such as transmission towers, cable conduit networks, earth control stations and associated equipment for space satellites.

4.9.2 Telecommunications Networks Engineering

The telecommunications networks engineer designs, directs installation and maintains fixed line, wireless, microwave networks, and backbone infrastructure. This includes selection of appropriate equipment, installation, administration and maintenance of telecommunications systems including control and transmission/receiving stations, circuit, network and complex electronic switching systems, installation and maintenance of fixed-line and wireless transmission, mobile phone networks, optical fiber cabling, IP networks, microwave transmission. This sub-specialty is in fact telecommunications hardware engineering but with focus on networks.

4.9.3 Broadcast Engineering

Radio and television broadcasting is perhaps the strongest global medium of communication and dissemination for news, advertising and entertainment, and much of it depends on telecommunications hardware: studio equipment, transmission equipment, mobile broadcasting units, etc. The Internet has greatly facilitated acquisition and dissemination of news and entertainment and more or less transformed the world into a global village. A telecommunications engineer can specialize in broadcast engineering which involves design and selection of appropriate equipment for radio and television broadcast stations, design of studios, transmitters, mobile broadcasting systems, direction of installation and maintenance of equipment. Electronic advertising has become a major tool in commerce, politics, the Internet, social media, etc., and telecommunications engineers are now specializing in this area. In addition to competence in telecommunications engineering, a good working knowledge of audio, video and lighting engineering, consumer targeting is also essential.

4.9.4   Career Opportunities in Telecommunications Engineering

The telecommunications industry is perhaps the most dynamic of all global economic sectors, and is taking over control of our daily lives: from the smart phone, Internet connectivity and dependence, to the explosion in the social media, news media, advertising, and entertainment. Telecommunications Engineering is considered as the backbone of every industry as it includes installation and maintenance of multiple types of network systems like optical fibres for Wi-Fi, basic telephone connections, and microwave transmission systems, among others. Considering the exponential growth rate of developments in the last two decades or so, it is not difficult to speculate on an even faster future growth rate of global dependence on telecommunication: local and international trade, Internet commerce, banking, social interaction, radio, television, education, entertainment, healthcare, utility deployment, and many more.

The number of current mobile phone users in the world is estimated at about five billion (around 80% of the global population) and all depend on providers, masts, control stations, satellites. Telecommunication devices rely on computer/control hardware networks to function and telecommunications engineers have a very wide range of career opportunities in network design, deployment, operation and maintenance; network integrity, security and administration; radio and television broadcasting; sound engineering; the entertainment industry; advertising; and many more. While the major employers in the telecommunications industry are the big telecommunications companies and broadcasting networks, opportunities also abound in all sectors of industry and commerce, hospitals, educational institutions which maintain computer hardware and networks. Telecommunications engineers also establish independent businesses that serve as contractors for the larger telecommunications companies, for example in installation of masts and cable networks, supply and installation of equipment, and maintenance of control stations.

4.10 NUCLEAR ENGINEERING

Atoms are the building blocks of every object in the universe. The nucleus (core) of each atom contains protons and neutrons held together by nuclear energy. The energy can be released by bombarding radioactive uranium atoms with neutrons. Split atoms (fission) release enormous energy which can be used to generate steam and produce electricity or power bombs deployed in military warfare. Nuclear energy can also be generated by combining two atoms to form a larger atom (fusion) which is the source of the Sun's enormous energy, and extensive research is ongoing to simulate the same reactions in the laboratory. Nuclear engineering developed in the nineteen thirties, largely motivated by its potential for military applications, ultimately leading to the first atomic bomb which was deployed in the second World War in the late nineteen forties. To date, the enormous potential for nuclear weaponry has been the primary force driving the development of nuclear energy. Apart from much more powerful nuclear weapons stored in some countries today, there are many nuclear-powered military ships and submarines that can operate on small quantities of fuel for months and years. However, peaceful applications of nuclear energy have also evolved over the years, notably in electric power generation, space exploration, medical diagnosis and treatment, non-destructive testing, etc. There are four basic specialties in nuclear engineering as shown in Figure 4.10. While some institutions offer first degree courses in nuclear reactor engineering, postgraduate programs are the most common and people who hold a first degree in physics, mathematics, electrical, computer, mechanical, materials or chemical engineering can easily retrain as nuclear engineers, specializing in different aspects. This later route is more common and offers more flexible employment opportunities outside nuclear reactor engineering.

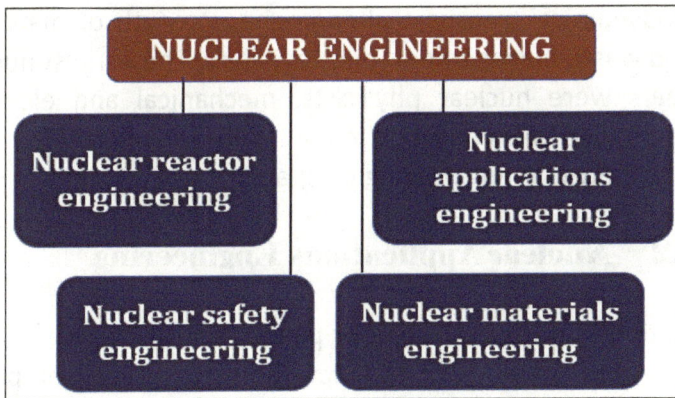

Figure 4.10. Sub-specialties in Nuclear Engineering.

Areas of specialization include nuclear reactor engineering, fuel management, nuclear applications engineering, nuclear materials engineering, radiation monitoring, safety and security. Nuclear engineering training programs offer core courses in all these areas after which the student takes additional courses related to specific areas of specialization.

4.10.1 Nuclear Reactor Engineering

The process of splitting or fusing atoms is very specialized and extremely dangerous, hence nuclear reactors are located deep underground in concrete structures to contain potentially harmful radiations. Nuclear reactor engineering involves design and operation of nuclear reactors, notably the reactor core, fuel rod assembly and handling, controlled and safe manipulation of fission/fusion reactions, selection of associated equipment, control of cooling systems and disposal of contaminated effluent water, design of radiation shields, design, deployment and operation of radiation and safety monitoring equipment and procedures, toxic and radioactive waste management, and decommissioning of nuclear reactors which have a lifecycle of around thirty years. A strong background in nuclear physics is a prerequisite, but a typical nuclear engineering program also includes basic engineering: engineering mechanics, engineering thermodynamics, transport phenomena, strength of materials, fuel and waste management, nuclear safety. The earliest nuclear engineers were nuclear physicists, mechanical and electrical engineers who retrained but many institutions now offer fully integrated nuclear engineering programs.

4.10.2   Nuclear Applications Engineering

Application of nuclear energy for peaceful purposes emerged in the nineteen fifties, with the development of nuclear power stations which use nuclear energy instead of coal or petroleum fuels to raise steam that drives turbines for power generation. This aspect of nuclear energy use involves high-energy nuclear and steam energy generation and utilization, managed by nuclear reactor engineers, mechanical and electrical engineers with competence in design, operation and maintenance of nuclear reactors, steam raising plants, steam turbines and electric alternators. Low-energy nuclear radiation applications have also extended to industry and medical practice, and a very wide range of radioisotopes-based equipment for industrial non-destructive testing and non-evasive medical diagnosis is now available. These include non-destructive testing machines for evaluation of materials' integrity in service and medical equipment which use radioactive isotopes, for example, use of ionizing radiation and radio-pharmaceutical markers: X-ray, computed tomography (CT), mammograms, molecular imaging (MI), magnetic resonance imaging (MRI) etc., and minimally invasive medical treatment (radiotherapy). The nuclear applications engineer has the expertise to design radio-diagnostic and treatment equipment and may also be involved in commissioning, maintenance and safe disposal of decommissioned (though still potently harmful) equipment. Nuclear applications engineers can specialize in specific fields of such as reactor design, materials testing, radio-biomedical investigation and treatment.

4.10.3   Nuclear Materials Engineering

Human exposure to nuclear radiation is very dangerous and could be lethal hence special materials and procedures are required to contain radiation from the fuel, the nuclear reaction and the reaction waste. The reactor is always installed underground and embedded in thick concrete to contain radiation leaks. Very few metals can resist radiation damage notably some special steel and zirconium alloys, concrete, ceramics and graphite which are used in different parts of a nuclear plant: fuel rod assembly framework and casing, reactor core, moderator rods, radiation shields, waste encasement, etc. Nuclear reactor cooling water systems contain corrosion residue which needs to be monitored by materials engineers to prevent plant shutdowns and decrease safety-related incidents. The nuclear materials engineer needs a strong materials science and engineering background and additional training in materials' behaviour in radioactive environment, materials selection for nuclear applications.

4.10.4   Nuclear Safety Engineering

Safety is a critical issue in nuclear engineering for many reasons: Human exposure to large doses of uranium radiation is extremely harmful especially when it comes from the reactor; spent reactor fuel remains hot and radioactive for decades; interruption of reactor cooling water even for a very short time can cause serious disaster as was the case with the three serious accidents discussed earlier; cooling water remains contaminated for decades and safe disposal is a major problem. Furthermore, radioactive waste currently being stockpiled around the world may remain potent and dangerous for hundreds of years and the nuclear safety engineer is well trained in best waste handling and management practices. The core responsibility of a nuclear safety engineer is to design safety measures and ensure compliance. This includes monitoring of radiation levels, overseeing the provision of appropriate radiation protection wear, regular accident drills, safe handling of nuclear materials and disposal of nuclear waste. Any science or engineering graduate can train as a nuclear safety engineer.

4.10.5 Career Opportunities in Nuclear Engineering

Nuclear engineering is highly specialized and the main career opportunities are in military applications, nuclear power generation and nuclear research. However, the scope of non-military, non-power use of radioactive materials is also growing very rapidly - radioisotopes are in common use in medical diagnosis and treatment, and in industry. Nuclear power plants, nuclear-powered ships, research and development account for around 60% of employment opportunities in nuclear engineering, and most employers offer training on site to new engineering recruits (most engineers can be retrained in one aspect of nuclear engineering or the other) in such areas as nuclear reactor operation, materials selection and handling, radiation shield management, operational and safety procedures and regulations. It should be noted however that opportunities are largely country-specific: there are only around 450 operating nuclear reactors in the world; around three quarters are in the United States, France, Japan and China, and most of the research and development opportunities are also in these countries. Nuclear engineering is considered a security asset and employees in nuclear establishments are usually screened. Also, while most medical and industrial radioactive equipment are designed by nuclear engineers and manufactured in only a few countries, operation and maintenance worldwide are carried out mostly by technicians who are trained mainly at the facilities of the equipment supplier. For these reasons, employment opportunities for nuclear engineers are low compared with other engineering specialties. It is advisable that students interested in nuclear engineering should first obtain a degree in another branch of engineering as a backup to broaden their employment opportunities.

4.11   ENERGY ENGINEERING

The world has access to several natural energy resources, known collectively as primary energy: solar, wind, fossil fuels (coal, oil, gas), biomass, hydro, geothermal, tidal energy. While they had all been used directly but minimally from Early times, processing is needed to produce large quantities of energy required by modern society. Energy from coal, oil, gas, solar, geothermal, biomass, nuclear fission is used to raise steam for electricity generation; fluid energy (hydro, wind, tidal) turns turbines to produce electricity; refineries produce gasoline, diesel oil, aviation fuel from oil, coal, biomass; hydrogen is produced from coal, natural gas or water to drive gas turbines for electricity production (Figure 4.11). Different aspects of energy engineering belong to different engineering specializations, notably geological, mechanical, electrical, chemical, nuclear and control engineering. However, in recent times, a new specialization has been emerging which brings together the different skills into a new area of specialization: energy engineering. A typical course includes all areas relevant to energy discussed already under the different specializations, from prospecting for primary energy to different production, refining and conversion processes. The basic course establishes a strong foundation in basic engineering subjects, energy production, conversion and distribution technologies, energy efficiency in production and utilization, energy services, facility management, plant engineering, environmental compliance, sustainable energy and renewable energy technologies. The student can then specialize in one of several main areas: petroleum, green energy, geothermal energy, or nuclear energy engineering.

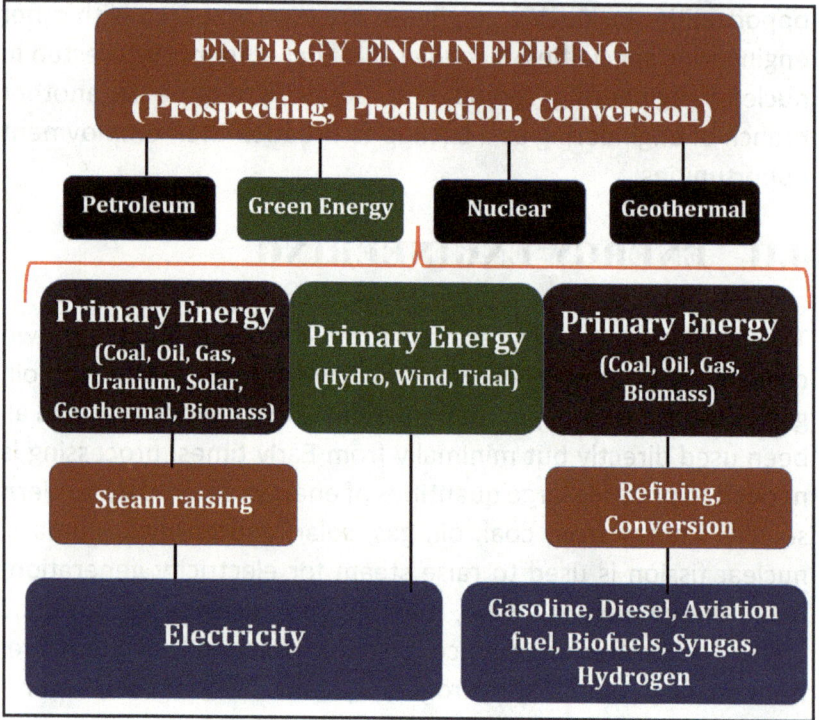

Figure 4.11. Sub-specialties in Energy Engineering.

4.11.1   Career Opportunities in Energy Engineering

Energy engineering is an emerging engineering discipline and employment potentials are very wide: electric power generation, fuel energy prospecting, production, refining/processing, storage, transportation, distribution, and utilization efficiency, environmental compliance, facility management, and many more. In spite of the negative environmental impact, fossil fuels still supply around 80% of global requirements of converted energy. Production conversion and distribution still provide the large majority of employment opportunities in many countries. However, increasing environment concerns are promoting cleaner energy (green energy) technologies, notably solar and wind. However, although growth particularly in the last decade has been phenomenal, these technologies are starting from a very low base, jointly accounting for less than around 6% of the global primary energy use in 2020. The demand for clean energy engineers is growing very rapidly, particularly in Europe, North America and China, and probably outstrips production already. However, graduates of other engineering specialties that have significant energy technology content can retrain and function well in green energy technology enterprises.

4.12   NANO ENGINEERING

Nano science/technology/engineering involves manipulation of biological and physical materials at the sub-micron level or nano scale which is about 1-100 nanometers (one nano or nm is one billionth or 10^{-9} of a meter). It is difficult to imagine just how small a nano is, but a single human hair or a sheet of paper is 80,000-100,000 nanometers thick (10^{-4} of a meter). Properties of materials at nano levels such as melting point, fluorescence, electrical conductivity, magnetic permeability, and chemical reactivity can be dramatically different compared with bulk material, largely because quantum effects rule the behavior and properties of particles at nano levels. This explains why nanoscale gold can appear red or purple depending on the size of the particle. Nano technology is a very complex field which cuts across many disciplines that operate on the nano scale: biology, physics, materials science, and the ability to work with and control materials on a nano scale is leading to a revolution in the materials world, notably in materials science and engineering and medicine. It is now possible to observe, characterize and manipulate materials on a nano scale, and create important products with size and properties never before envisioned.

Although nano materials have been in use since medieval times (for example, nano scale gold was used extensively in staining glass Figure 4.12a), the real revolution came with the invention and development of the electron microscope from the 1930s which now makes it possible to see and work with nano materials. With so many advances over the last decade, nano technology has become a new, major frontier of engineering, creating endless possibilities, with nano products already featuring prominently in manufacturing, information technologies, energy technologies, biomedicine, microfluidics, robotics, heat transfer and storage, computational modeling, building and household materials, textiles, cosmetics, food, environmental technologies, medicine, pharmaceutical products, etc. Carbon-based nano materials known for high strength-to-weight, and good conductivity are featuring in many areas of auto and aerospace applications. Nano engineering is also one of the most interdisciplinary of the sciences, requiring knowledge of mechanical engineering, chemical engineering, electrical engineering, biology, physics, photonics, and materials science. Nano technology offers vast amounts of enhancement in virtually every area of engineering. The two major areas of nano engineering are shown in Figure 4.12b.

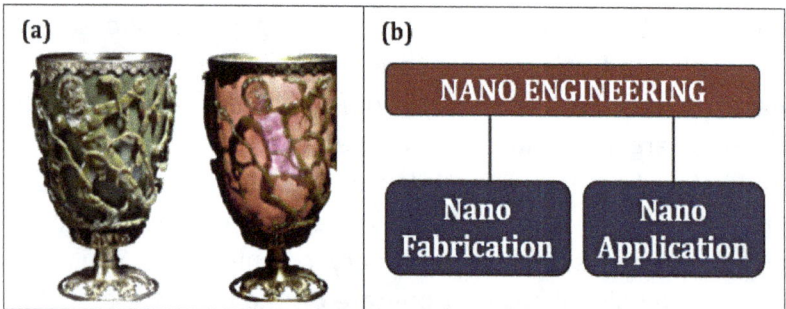

Figure 4.12. (a) Fourth century AD Roman Lycurgus nano-gold cup changes color depending on direction of light. (b) Sub-specialties in Nano Engineering.

4.12.1   Nano Fabrication Engineering

Nano fabrication engineering is the aspect of nano technology that deals with the manipulation of materials on nano scale to create and fabricate products that are vital to the advancement of science, technology, engineering and medicine. Using basic biochemical processes at the atomic or molecular level, the natural interaction of molecules can be manipulated to assemble and grow in a desired manner in order to achieve various arrangements that confer specific properties on the product. The precursor material may be a metal, metal oxide or carbon.

The production process of nano materials is critical to the ultimate engineering properties of the product. The advanced structural manipulation at molecular or even atomic levels under highly controlled conditions and environment confers unusual and extraordinary mechanical, electrical, chemical, optical and magnetic properties compared with same materials produced by the traditional ore-to-product process. For example, silicon produced by the traditional method of reducing silica with carbon at very high temperatures is very hard and brittle. However, silicon nano wire produced by nano technology is ductile and around a thousand times stronger than steel. Also, when embedded in organics (both conductive and nonconductive) the product combines strength with flexibility, ductility, and enhanced conductivity, and is finding increasing use in organic-based electronics, flexible electronics, wearable electronics, etc. Silicon nano wires are also combined with ceramics to produce composites with improved mechanical properties.

Materials produced by nano engineering and technology include nano carbons, nano polymers (dendrimers) and nano composites. These materials are finding increasing applications in industry and biomedical engineering. For example, downsizing the computer from room-filling box to a laptop, or the invention of the multi-purpose smart phone (which is in fact a minicomputer) was made possible because of nano engineering which can etch millions of electronic circuits on silicon wafer substrate the size of a coin (semiconductor/integrated circuit/chip). This new technology has made it possible to tailor the structures of materials at very small scales to achieve a desired set of properties, and the scope of applications is extremely wide: materials science and engineering, biomedical engineering, information technology, control engineering, textile technology, telecommunications, transportation and aerospace engineering, military engineering, energy, etc.

4.12.2   Nano Applications Engineering

Nano applications engineers specialize in developing new areas of application for nano materials in industry, medicine, weaponry, etc. Data storage systems are becoming increasingly miniaturized: a 2-gigabyte computer hard disc of the 1980s weighed around 30 kilograms and cost around a hundred thousand dollars, compared with today's USB drive of same storage capacity that weighs only a few grams and costs a few dollars and, with developments in nano technology, further miniaturization is only a matter of time. It is now possible to double the strength of a product or improve resistance to corrosion by combining two nano materials into composites. For example, medical implants (embedded in aggressive body fluids) and extremely lightweight bullet-proof vests are now being fabricated from carbon nano tubes and nano composites, and other potential areas of applications are very wide.

Nano particles are being used extensively in drug delivery, for example, chemotherapy drug carriers have greatly enhanced effectiveness in cancer treatment and minimized damage to healthy cells. New lightweight, stronger, lighter, more durable materials are emerging almost daily for applications ranging from military to domestic products. Carbon nano tubes are now being fabricated into lightweight military body armor, micro sieves, etc., and nano particles are used extensively in conditioning fabrics to improve resistance to wrinkling, staining and bacterial growth. Composite materials formed by embedding carbon nano tubes in metal or plastic matrix are now favorite materials for a wide variety of products that require high strength-to-weight ratio, such as aerospace components, sports equipment, military gear, laptop and cell phone cases, etc. The fuselage and many components of some of the most modern aircraft are now fabricated from nano-scale aluminum-carbon fiber composites, leading to around 30% reduction in weight and significant reduction in fuel consumption. For example, around 60% of the components of the world's biggest airplanes, Boeing 787 and Airbus 380 aircraft are fabricated from aluminium-carbon fibre composites. Nano engineering and technology is revolutionizing lithium-ion battery technologies which now power modern aircraft, electric vehicles, laptops, cell phones, etc., and has great potential in solar power storage. Nano materials are also finding increasing use in display and flexible electronics, data storage, nano structured semiconductor devices, thin film solar cells, energy storage systems, nano agriculture, nano water purification, and nano medicine, nano electro mechanical systems (NEMS), biomarkers, biomedical implants, nano electronic biosensors, diagnosis and treatment of cancers, molecular nano technology, etc.).

4.12.3   Career Opportunities in Nano Engineering

Nano materials are considered materials of the future and nano engineering is the fastest growing research area in materials science and engineering. Major areas include technologies for the production of single crystals, nano powders, ceramic and polymer nano composites, and high-technology materials which have emerged in recent times include nano-crystal silicon which is around a thousand times stronger and tougher than steel, nano wires, tubes, plates that are many times stronger, tougher and harder than equivalent conventional materials. Training in nano technology is usually at postgraduate level and a good degree in physics, chemistry, biological sciences, electrical and electronic engineering, or materials science and engineering is a suitable prerequisite. Nano engineers work mostly in research and development but also in industries that produce advanced materials or manufacture advanced electronic and control equipment, semiconductors, smaller and more efficient processors, memory chips, data storage systems, energy storage systems, medical implants, aerospace components, etc. Many nano engineers also work in medical and pharmaceutical research, creating and manipulating micro-molecules that can fix various health problems. The field of environmental monitoring is also becoming an attractive area for nano engineering since many of the toxic pollutants are made up of extremely small elements and particles which are only visible under very high magnification.

4.13   BIOMEDICAL ENGINEERING

Biomedical engineering is the application of engineering principles and design concepts to biology and medicine for therapeutic, diagnostic, monitoring, reconstruction healthcare. It is an evolving discipline and seeks to integrate human biology with relevant areas of engineering, notably mechanical, electrical, chemical, materials science and engineering, and computer science and engineering. As explained earlier, the human body and engineering devices have a lot in common and many basic engineering principles also apply to the human body. For example, fluid dynamics is a powerful subject in many engineering disciplines but is also applicable to cardiovascular blood flow to analyze unsteady blood flow, flow through heart valves, blood flow and cardiac chamber fluid-structure interaction, and flow related blood damage. Many standard theories of engineering mechanics are also applicable to human joints. A typical curriculum would include basic foundation engineering subjects, bio mechanics, tissue mechanics, cell and molecular biology, bio materials, bio and nano technology, statistical and numerical techniques, fabrication of medical devices and sensors, biomedical micro-controllers, imaging, biomedical design.

Biomedical engineering differs from other engineering disciplines that have an influence on human health in that biomedical engineers use and apply an intimate knowledge of modern biological principles in their engineering design process. There are many sub disciplines but the most developed are tissue and stem cell engineering, implant engineering, and medical imaging. Biomedical engineers design and develop active and passive medical devices, biocompatible implantable devices such as pacemakers, heart valve replacement, coronary stents, artificial hips, orthopedic implants, artificial body part replacement (prosthetic/bionic limbs, eyes, etc.), biomedical sensors, dental products, and ambulatory devices. Biomedical engineers are working towards artificial re-creation of human organs, notably the heart, lungs, kidneys, aiding in transplants and helping millions around the world who live with defective organs. Much of the work of biomedical engineers consists of research and development spanning a very wide range of subjects: regenerative cell and tissue engineering to regenerate diseased or injured tissue; genetic engineering; rehabilitation engineering; methods for repairing and replacing damaged or diseased organs, measuring the internal structures of the human body in health and in disease; development of new diagnostic tools; etc. Clinical engineers design and develop imaging equipment (MRIs, EKG/ECGs, etc.) and work to ensure that medical equipment is safe, reliable, and well maintained for use in clinical settings.

Research is extensive and intensive in finding new materials for implants, neural engineering, drug engineering (drug delivery and targeting). The emergence of chitin and chitosan as very versatile and promising biomaterials was discussed briefly in Section 4.5.4.

Nano technology is providing new micro encapsulation techniques for fabrication of complex nano carriers for the delivery of drugs, biologics and vaccines. Chitin and chitosan are used in a wide range of biomedical applications such as tissue engineering (repair and regeneration), drug and gene delivery, wound healing, and stem cell technology (Berada, 2021). These natural biopolymers can be easily processed into various products, including hydrogels, membranes, nanofibers, beads, micro/nanoparticles, scaffolds, and sponges. The most important characteristics which have made them ideal candidates to fabricate polymeric tissue scaffolds are: high porosity; biodegradability; predictable degradation rate; structural integrity; non-toxicity to cells; and biocompatibility. In the form of nanoparticles, chitosan can be appropriately formulated to alter the physicochemical properties of drugs and enhance absorption. Chitin-based nanoparticles are being loaded with various drugs and used as controlled oral delivery systems for active pharmaceutical ingredients (API), notably peptide, protein, vaccine. Chitosan-coated semiconductor nanocrystals are used for bioimaging in cancer diagnosis. Composites of chitosan and calcium phosphates hold promise as a substitute in restoration of damaged hard tissue (bone). Chitosan film impregnated with carbon dots are used in wound dressing and are being tested as trackers of biological processes inside human cells. Chitosan and its functionalized form, as well as its nanoparticle composites have been reported and widely used as good anti-corrosion compounds for different metal/medium systems in preference to the more common inorganic coatings because of its adsorption capability on the metal surface via electrostatic attraction and/or chemical bond formation, thereby providing much better corrosion mitigation. Chitosan, its derivatives, and the complex formation with other substances have been used for applications in filtration and membrane separation processes.

4.13.1 Career Opportunities in Biomedical Engineering

Career paths in biomedical engineering are very wide, considering the huge size and diversity of the global healthcare sector. Medical diagnostics triple in market value each year. Revolutionary advances in medical imaging and medical diagnostics are changing the way medicine is practiced all over the world. Biomedical engineering is one of the most researched areas of technology and opportunities abound in research and development. New medical devices coming from the research laboratories of biomedical engineers around the world, have completely altered the manner by which disease and trauma are diagnosed and treated by physicians, extending the quality and length of human life. Virtually every hospital requires some products of biomedical engineering and operates extensive associated medical equipment. Apart from research and development, opportunities also abound in marketing of biomedical equipment and consumables, and maintenance services.

4.14 ARCHITECTURAL/BUILDING ENGINEERING

Architecture is a creative art applied to the design of buildings, bridges, and other structures. Architects are not engineers and produce a design expressed in working drawings, to be actualized by other specialists. They also supervise their projects to ensure compliance. They need other engineers (structural, mechanical, electrical) to complete a design. The actual construction requires a team comprising civil, structural, mechanical, electrical and environmental engineers. Architectural engineering combines the art and engineering principles of the design and construction of buildings and building systems. It is an evolving interdisciplinary engineering specialty which promotes a systems approach to building design and construction by integrating the different specialties of planning, design, construction, operations and management of buildings, and building environmental systems. In effect, an architectural engineer is trained to do work that would normally require a team of four or five different engineering specialties. A typical curriculum provides a strong foundation in creative building art design, basic principles of civil engineering, structural engineering, computer-aided design, air conditioning, electric power systems, plumbing, lighting services, acoustics, fire protection, energy conservation, construction management. In some university programs, students can specialize by concentrating on one or two areas while in others, they can receive a generalist architectural or building engineering degree.

Building engineering is a specialized aspect of civil or architectural engineering. A building engineer is involved in planning, designing, constructing, operating, maintaining, supervising and renovating building structures. This includes airports, bridges, channels, dams, harbors, irrigation projects, pipelines, power plants, roads, railroads and water and sewage systems.

4.14.1   Career Opportunities in Architectural/Building Engineering

Graduates of architectural engineering are widely considered to be creative systems engineers, with formal training in creativity and design through architectural design studios married with a solid engineering education that greatly extends their skill sets. Buildings are vital essentials for any public or private enterprise, and building engineers have a very wide scope of employment opportunities. Like other engineers, architectural and building engineers are not actual 'doers' but they plan, manage and coordinate the work of engineering technicians who do the construction. The World population is growing fast and provision of adequate housing is a prime challenge in most countries.

The middle class is expanding, and well-built housing in a conducive environment tops the list of priorities of many families. Because people spend 86% of their time indoors, architectural/building engineers concentrate on indoor building environments that prioritize the human condition and well-being of society. Aging building infrastructure is being replaced or remodeled, new settlements and cities are being built. Civil engineers are already in very high demand but architectural and building engineers are emerging as the preferred choice because of their versatility which can reduce building construction costs significantly. The employment opportunities for architectural and building engineers are endless and wide-ranging. Graduates regularly accept job offers from architectural and construction engineering firms, consulting engineering firms, real estate developers, building equipment designers, manufacturers, designers and producers of building materials and products, facilities engineering and management groups, building owners, specialty contractors, forensic engineering consultants, building technology consultants, software developers, contractors, and construction managers. Opportunities to own building and contracting businesses are also high.

4.15 GEOLOGICAL/GEOTECHNICAL ENGINEERING

Geologists study the Earth, its physical environment, earth materials, and natural resources. Basically, they are scientists, they have the skills to locate mineral resources below ground, below ocean and earth's surface, groundwater resources, geothermal energy, but geological engineers combine science with engineering to develop and recover the world's natural resources for the use of mankind. There is a significant overlap with mining engineering which is a sub-discipline of mechanical engineering, since both involve the design and construction of mineral recovery systems. This requires characterization of the geophysics and geochemistry of the deposit, selection of the appropriate recovery process, design of foundations, support structures, product transportation systems, selection of the appropriate machinery, mine safety/mitigation of geologic hazards, and environmental engineering. While mining engineers focus on solid minerals, geological engineering combines mining with civil, petroleum and environmental engineering. Geological engineering is a relatively rare degree program and graduates have similar employment opportunities as mining engineers in a variety of industries including mineral resource extraction, petroleum exploration, production and service, hazard mitigation, engineering consulting, water resource development and protection, construction, environmental consulting. However, they have an edge in petroleum prospecting and production because of their broader skill sets.

4.16   SPORTS/AEROBICS ENGINEERING

Sports engineering is usually a postgraduate degree designed for graduates of mechanical engineering to give them a better understanding of sports equipment design and production. However, the scope of the field is not limited to only mechanical engineering, and has been emerging as a separate and distinct engineering sub-discipline. A typical degree program in sports engineering has a set of core courses: engineering mechanics (kinematics, statics, dynamics), biomechanics, materials science, design, production and maintenance of sports buildings and facilities, performance measurement, analysis, and athletic feedback systems. The field has a significant overlap with other types of science and engineering, notably (physics, mechanical, civil, electrical, materials science and engineering), and many practitioners hold degrees in one of those fields rather than in sports engineering specifically. However, some universities offer programs at either the undergraduate or graduate level. Also, some mechanical engineering degree programs offer specialization in sports engineering.

4.17 MILITARY ENGINEERING

The metallurgy of all the metals of antiquity (iron, copper, gold, tin, lead) derived from the Early Middle East, originally on small scale to produce the metals for making ornamental objects but iron, copper, bronze, brass (copper-tin and copper-zinc alloys respectively) soon emerged as the prime metals for military warfare which required construction of weapons, armors, shelters, roads, bridges, wagons, and so on, the humble beginning of military engineering, the oldest of the engineering specialties, and the precursor of civil engineering which evolved when the same technologies were adapted in civilian context. Modern military engineering is the art and practice of designing and building military works and of building and maintaining lines of military transport and communications. It comprises three basic specialties: combat/battlefield engineering which involves provision of tactical engineering on the battlefield: creation of physical cover, bridges, trenches, laying of mines, demolition of structures, etc. The second specialty is strategic support involving construction of support services such as communication infrastructure, camps, airfields, airports, port facilities, roads, bridges, hospitals, supply storage/distribution systems, troop movements, etc. The third specialty area is ancillary support such as provision and distribution of military intelligence to combat units and tactical groups, disposal of unexploded bombs, mines, warheads, maintenance of military equipment, general management of military facilities, etc. The military in most parts of the world tend to be eager to recruit aspiring engineers and usually fund in-service engineering training, allowing specializations in such strategic areas as construction, equipment/automobile/aircraft maintenance, radar technology, missile technology, all of which open up many post-service employment opportunities.

4.18   FORENSIC ENGINEERING

As discussed earlier, there are no perfect engineering designs and there are many factors which can cause even a perfect design to fail in service, in particular, the environment in which the product is located or operates and unforeseen stressors such as extreme weather or earth movements. Forensic engineering involves detailed investigation of failures and accidents ranging from serviceability to catastrophic, to determine the causes which may be a design flaw, materials failure, structural failure, overload, lapses in deployment and servicing, or a combination of factors. Apart from providing information which helps improve product and reduce chances of future failure, the information may be required in respect of civil or criminal investigations. While investigation may focus on a particular cause, few serious accidents are attributable to a single cause and may involve different experts: engineers with specializations relevant to the product, manufacturers (structural, mechanical, materials, electrical), regulatory bodies, etc. Systems engineers can specialize in forensic engineering and coordinate the work of the different engineers.

5   Women in Engineering: Yes We Can !

Engineering is gender neutral

5.1   INTRODUCTION

Engineering is part of the family of STEM (science, technology, engineering, mathematics). A STEM qualification in the United States places the holder in the top ten percent of the job market and a typical STEM worker earns two-thirds more than those employed in other fields, according to Pew Research Center. However, some of the highest-earning STEM occupations, such as computer science and engineering, have the lowest percentages of women workers. The reasons for the low footprint of women in STEM jobs are many and varied but some of the most prominent are discussed in the following sections.

5.2   THE STEM GAP

Girls and women are systematically tracked away from science and math throughout their educations, limiting their training and options to go into these fields as adults. Women make up only 28% of the workforce in science, technology, engineering and math (STEM), and men vastly outnumber women majoring in most STEM fields in college (AAUW, 2021). Women constitute 48% of biological scientists, 43% of chemists and materials scientists, 26% of computer and mathematical occupations, and only 16% of engineers and architects. The gender gaps are particularly high in some of the fastest-growing and highest-paid jobs of the future, like computer science and engineering (Figure 5.1).

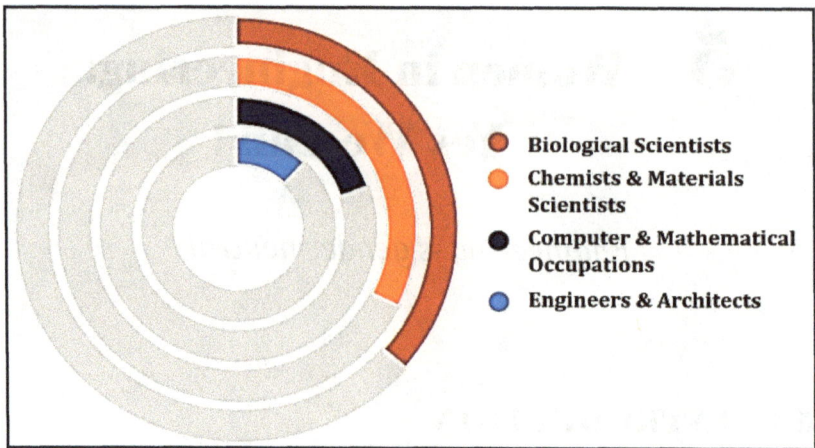

Figure 5.1. Women in STEM occupations in the United States. *(Source: U.S. Bureau of Labor Statistics).*

5.3   WHY SO FEW?

Women are few in STEM occupations because, from birth, they are made to believe that they are inadequate for many reasons discussed below.

5.3.1   Gender Stereotypes

Traditionally, engineering has been a male-dominated profession, just like the military, and there are many stereotypes out there that tend to discourage women from choosing a career in engineering. In many traditions, women are discouraged in many ways from receiving education, and when they have the opportunity, they are made to believe that mathematics and science are beyond their capabilities. One notable variant is the old Soviet model: women were encouraged to pursue careers in the sciences and medicine, but not engineering which was considered to be an exclusive preserve of men. Another myth is that engineering is too rough for women, but this arises from the confusion about what engineers do and the distinction between engineers, technologists, technicians and artisans (see Chapter 1.4). An engineer is hardly ever involved in physical processes of production, or equipment installation, operation, and maintenance. The work of the engineer is largely intellectual: design, analysis, project, process and product development, operations planning and supervision, systems design, development, deployment, operation and supervision, and the increasingly digital world has revolutionized and greatly complemented the way engineers do their work. One major attraction to engineering is the non-routine nature of work: the engineer is constantly faced with problems, issues and challenges that bring out the best of creativity instincts and competence, and these gifts are gender-neutral.

5.3.2   Male-Dominated Cultures

Males tend to want to preserve the cultures they dominate, and STEM is a prominent example. Because there are so few women in STEM employment, men largely control employment opportunities and tend to perpetuate the status quo: the inflexible and exclusionary male domination. Women who succeed in breaking the 'glass ceiling' usually find that they are rated and remunerated at much lower levels than their contemporaries. Men in USA STEM jobs earn 40% more than women on average. This situation tends to discourage women from aspiring to STEM careers.

5.3.3   STEM Anxiety

As early as pre-school, girls are made to believe that STEM subjects, especially math, are difficult and beyond their competence. This myth is passed on through generations and girls tend to lose confidence in their ability. However, research has so far identified no innate cognitive biological differences between men and women in math. The negative attitude of parents is also a significant contributory factor because they take as normal the poor performance of their girls in math whereas they go to great lengths to help boys who are struggling in the subject, such as paying for supplementary tutorship. This explains why boys from higher income families tend to perform significantly higher in math than other boys and girls in the same school. It is worthy of note that girls do better in elementary school math than boys from the same low income group (AAUW, 2021). Also, experience has shown that women who defy the norms are as competent, and are even out-performing men in the physical sciences and engineering at college level, and those that succeed in breaking down the glass ceiling in the employment market are hardly ever found wanting in terms of competence and effectiveness

5.3.4   Paucity of Role Models

The fact that there are relatively few women in STEM employment means there is a paucity of role models who can inspire young girls to study STEM. There are limited examples in families, communities or the media, hence there is little motivation anywhere. The story of Marie Curie should be a source of inspiration to young STEM-oriented girls. Marie was born in Poland in 1867, the daughter of a school teacher. She studied science in France and obtained her doctoral degree in physics and mathematical sciences in 1903. She was awarded the world's most prestigious STEM prize: the Nobel Prize for Physics jointly with her husband Pierre in 1903 (three-quarters of the award was for her). This was as a result of their ground-breaking work on the discovery and isolation of radioactive polonium and radium from pitchblende mineral. In 1911 she received a second Nobel Prize, this time in chemistry for her work on the isolation of pure radium metal, and became the first woman to win a Nobel Prize, the only woman in history to ever win it twice, and the only woman to ever win a Nobel Prize in two different fields of science, among many other firsts. Although Marie Curie died in 1934 from leukemia caused by four decades of exposure to radioactive substances, she pioneered medical applications of radioactivity which is now widely applied in the treatment of cancers, many other medical conditions, and in industrial non-destructive testing. It is not a surprise therefore that she is regarded as not only one of the greatest scientists to have ever existed but also a symbol of female involvement in science, at a time when it was something that was not as common. It is also noteworthy that two women won the Nobel prize in physics and chemistry respectively in 2018; three women won the prize in 2020, one in physics and two in genetics. Katalin Kariko won the 2023 Nobel prize for physiology/medicine (with Drew Weissman) for her work on effective mRNA which led to the development of most COVID-19 vaccines.

In spite of oppressive institutions, cultural norms and other gender barriers in STEM fields, many women have persevered and contributed immensely to the advancement of science and technology. The first computer language which became known as COBOL was developed by Grace Hopper, a female American computer scientist in 1953, specifically for the first commercial computer, the UNIVAC. Nancy Grace Roman, the world famous astrophysicist made discoveries about the compositions of stars that had implications for the evolution of our Milky Way galaxy. Her pioneering work with the Untied States National Aeronautics and Space Administration (NASA) from 1959 led to the development of the Hubble Space Telescope, the world's most scientifically revolutionary, most powerful and most productive space telescope of all time. Apart from her universal recognition as the 'mother' of Hubble Telescope, NASA has recently named the next generation of space telescopes after her: Nancy Grace Roman Space Telescope.

Many women have occupied or are currently occupying leadership positions in industry, commerce, consulting, research and the academia all over the world. These include Mary Teresa Barra the Chairwoman and CEO of General Motors Company, the first female CEO of a major global automaker; Marilyn Hewson, CEO of Lockheed Martin Corporation, an American global aerospace, defense, security and advanced technologies company with worldwide interests; Safra Catz who heads Oracle Corporation an American multinational computer technology corporation and one of the world leaders in marketing database software and technology, cloud engineered systems, and enterprise software product; and Indra Nooyi, the CEO of Pepsi Cola. Hewlett Packard, one of the world's leading manufacturers of computers and business machines has been headed thrice by women in the last twenty years. Gwynne Shotwell is an aerospace engineer and President/Chief Operating Officer of SpaceX, one of the world's leading manufacturers of advanced rockets and spacecraft. Karen Matthys is the Executive Director of the Institute for Computational and Mathematical Engineering at Stanford University, U.S.A. Laura Boccanfuso is the Founder and CEO of Van Robotics.

Undoubtedly, women have proved that they can excel in STEM careers, but progress has been slow and women still constitute less than a third of all scientists and engineers in the United States which is one of the leading countries in gender emancipation, and the numbers vary very widely depending on the profession. According to a recent research, only 15.9% of engineers in the workforce in 2022 were women, probably because many of the women who graduated were foreign students who have returned to their countries, while others may have opted for non-engineering employment. Around a third of those who were employed in engineering were in computer science and engineering. The research also revealed that female engineers were earning around 10% less than their male counterparts and 61% reported that they had to prove themselves repeatedly to get the same level of respect and recognition as their male colleagues. These statistics from the most advanced economy in the world show that there is still a lot to do to fully integrate women into the engineering profession.

The Space Exploration Programs began in both the United States and Soviet Union in the late nineteen fifties and both men and women were recruited for training. Soviet cosmonaut Valentina Tereshkova was the first woman in space in 1963 but the second woman Svetlana Savitskaya only followed in 1982. All U.S.

astronauts sent to space for the first twenty years of the space program were male, perhaps because astronauts were required to be military test pilots, a profession that was not open to women at the time. The U. S. National Aeronautic and Space Agency (NASA) recruited the first set of female astronauts in 1978 in response to the new anti-discrimination laws. Even though women are now involved in many areas of space missions and technologies, design specifications for space shuttles' suites clearly target male astronauts. Kathryn Sullivan became the first U. S. Female astronaut to walk in space in 1984 yet, for the next thirty years, space suits were being designed to fit men, preventing women from being assigned to space walks because there were no suitable suits. The recent Women's History Month was to have been marked by the first all-female space walk but one of the two women had to be replaced because only one space suite was available in their size.

According to Meg Ury, Professor of Physics and Astronomy at Yale University, *"This is a disappointing reminder of how gender bias shapes our world".* According to her, *"Women who work in mostly male professions face obstacles that simply don't confront men in the same situations. An all-woman astronaut crew would have been a special boost to the morale of many women who are in science and technology professions and those aspiring. Difference is good, even great. It's known to lead to greater innovation. But sometimes sameness is hugely innovative too".* However, It is noteworthy that women scientists and engineers are playing increasingly prominent roles in the organization. For the first time ever and in a male dominated field, women are in charge of three out of four science divisions at NASA: Earth Science, Helio-physics and Planetary Science. These women are following the footsteps of many like Dorothy Vaughan, a respected mathematician who headed the National Advisory Committee for Aeronautics (NACA), the predecessor of NASA in 1948; Mary Jackson, the first African American mathematician and aerospace engineer who joined NASA in 1958 and rose to the highest engineering position; Ellen Stofan, a planetary geologist who serves as the Chief Scientist; Ellen Ochoa, four-time space shuttle astronaut and former director of Johnson Space Center. Hundreds of other STEM women are currently playing prominent and critical roles in many of NASA's research and development projects (see nasa.gov/stem/womenstem). Recently, the organization announced the Artemis project which plans to send the next set of astronauts (male and female) to the moon within the decade. The last moon trip over fifty years ago was named 'Apollo' the god of Greek deity mythology whose twin sister Artemis was the 'goddess of the moon.' Women scientists and engineers are playing very prominent roles on the project. In fact, the lunar landing spacecraft that will land the next man and first woman on the surface of the moon as soon as 2024 will be designed and built at the Marshal Space Center, and the project will be managed by a woman: Lisa Watson-Morgan, a three-decade veteran NASA engineer.

5.3.5   The STEMinist Movement

A lot of women have made and are making indelible marks in the field of STEM but there is paucity of information. A new non-profit organization: the STEMinist Movement has emerged comprising successful women who work in Science, Technology, Engineering, and Math (STEM) and who advocate for women in STEM fields, encouraging them to pursue these types of careers. They promote STEM interests in middle school girls by exposing them to different up-and-coming fields and engaging in discussions and holding workshops with the hope of empowering the next generation of innovators. However the reach is currently limited, mostly in the United States.

INSPIRATIONAL QUOTES: WOMEN IN STEM

Marie Curie Inspirational Quotes

(Marie Curie was a physicist, chemist, and winner of the 1903 Nobel Prize in Physics and the 1911 Nobel Prize in Chemistry)

"I was taught that the way of progress was neither swift nor easy."

"Life is not easy for any of us. But what of that? We must have perseverance and above all confidence in ourselves. We must believe that we are gifted for something and that this thing at whatever cost must be attained."

"First principle: never to let one's self be beaten down by persons or by events."

"I have no dress except the one I wear every day. If you are going to be kind enough to give me one, please let it be practical and dark so that I can put it on afterwards to go to the laboratory."

Other Inspirational quotes

"Science is not a boy's game, it's not a girl's game. It's everyone's game. It's about where we are and where we're going.
– Nichelle Nichols

"Stereotypical gender roles are present in many fields outside of medicine, including the office, household, and politics. The idea of "the woman's role" in these fields has become engrained in our society. Simply put, the man is the physician, the boss, or the breadwinner and the woman is the nurse, the secretary, or the   homemaker……..Defy the stereotypes !!" – Brittany Ladson, Medical doctor

"Don't let anyone rob you of your imagination, your creativity, or your curiosity."
– Mae Jemison, physicist and astronaut

"Science and everyday life cannot and should not be separated."
– Rosalind Franklin, chemist and X-ray crystallographer

"If you know you are on the right track, if you have this inner knowledge,   then nobody can turn you off... no matter what they say."
– Barbara McClintock, cytogeneticist and winner of the
1983 Nobel Prize in Physiology or Medicine

"I hadn't been aware that there were doors closed to me until I started knocking on them." – Gertrude B. Elion, biochemist, pharmacologist, and winner of the 1988 Nobel Prize in Physiology or Medicine

"With men you're expected to do well and get the support, but for females, you have to sacrifice something in a different way from men... I think it's very important for other woman to see that
I have had   success."
– May-Britt Moser, neurologist and winner of the
2014 Nobel Prize in Physiology or Medicine

"Life need not be easy, provided only that it is not empty."
— Lise Meitner, physicist

6 References and Bibliography

AAUW (2021). American Association of University Women.
  aauw.org/resources/research/the-stem-gap/
Afonja. A. A. (2020) "Mitigating Climate Change: The Power of We the
  People." SineliBooks.
Basalla, George (1988) The Evolution of Technology, Cambridge Uni-
  versity Press, Cambridge
Bath University (2018). "Engineering programs". bath.ac.uk.
Berada, M. (Ed) (2021), "Chitin and Chitosan: Physicochemical Proper-
  ties and Industrial Applications." intechopen.com
Berkeley (2019). "Career Field - Engineering and Computer Science".
  Career.berkeley.edu/Engineering.
  (Accessed 1/16/2019).
Cardwell Donald (1994) The Fontana History of Technology, Fontara
  Press, London Daumas
CCSU (2018). Engineering Disciplines. ccsu.edu.
  (Accessed 11/5/2018).
Educating Engineers (2019). educatingengineers.com
  (Accessed 2/9/2019).
Elieh-Ali-Komi, D. and M. Hamblin (2016). "Chitin and Chitosan: Pro-
  duction and Application of Versatile Biomedical Nanomaterials." Int
  J Adv Res (Indore). 2016 Mar; 4(3): 411–427.
G.E.M. Environmental (2018). " Explore STEM - The working Realm of
  Engineering". Gemenvironmental.org.
Lawlor, R. (ED) (2013). "Engineering in Society. Royal Academy of Engi-
  neering.
ICL (2019). "Engineering programs". Imperial College, London.
  imperial.ac.uk.
MIT (2018). "Engineering programs" mit.edu.
  (Accessed 07/30/2018).
Primack, B, *et.al* (2017). "Social Media Use and Perceived Social Isola-
  tion Among Young Adults in the U.S." Research Article Volume 53,
  Issue 1, P 1-8. Ajponline.org.
Stanford (2018), "Engineering programs". stanford.edu.
SWE (2019). "Women in Engineering". Society of Women Engineers,
  swe.org. (Accessed 2/9/2019).
Vanderbilt (2019). Summary of Engineering Disciplines.
  engineering.vanderbilt.edu. (Accessed 1/18/2019).